ARDUINO FOR BEGINNERS

Step-by-Step Guide to Arduino
(Arduino Hardware & Software)

Simon Knight

© Copyright 2018 by Simon Knight - All rights reserved.

The following eBook is reproduced below with the goal of providing information that is as accurate and reliable as possible. There are no scenarios in which the publisher or the original author of this work can be in any fashion deemed liable for any hardship or damages that may befall them after undertaking information described herein.

The transmission, duplication or reproduction of any of the following work including specific information will be considered an illegal act irrespective of if it is done electronically or in print. This extends to creating a secondary or tertiary copy of the work or a recorded copy and is only allowed with the express written consent from the Publisher. All additional right reserved.

Table of Contents

Introduction 6

Chapter 1: Getting Started with Arduino 8

Chapter 2: Hardware and Tools 18

Chapter 3: Turn your Arduino into a Machine. 54

Chapter 4: C Language Basics and Functions 64

Chapter 5: Advanced Inputs, Outputs, and Sensors 78

Chapter 6: Sound 94

Chapter 7: Arduino Shields 102

Chapter 8: Projects 114

Chapter 9: Troubleshooting 130

Chapter 10: Make your Project 136

Conclusion 138

Introduction

Congratulations on downloading *Arduino* and thank you for doing so.

Arduino is an open-source platform that helps people to develop electronic projects. Arduino comprises both the hardware and software. Arduino comes with an IDE that you install in your computer to write and upload code to the physical board.

Nowadays, the level of popularity of Arduino has spread with many people wanting to build something. Unlike the traditional circuit boards, Arduino does not call for an independent piece of hardware so that it can install a new code on the board. The included USB cable is more than enough. In addition, the Arduino IDE uses a version of C++, making it simple for one to learn how to program it.

Have you ever asked yourself how the coffee machine can tell the temperature at which to heat the coffee? Well, the small brain behind the commercial coffee machine is the microcontroller.

It receives things an analog input from a thermistor and displays the temperature in a digital display. The same can be applied to a remote control of a car that accepts commands from a wireless controller.

This enables one to regulate motors that rotate the wheels. The Arduino platform consists of a programmable board that has inputs and outputs to allow your physical project to function.

This book contains several chapters that will help you develop an in-depth understanding of the Arduino board. You will learn how to use Arduino and develop projects using it.

This book will prepare you to get started with your journey to develop electronic projects driven by the microcontroller. The different chapters will teach you both electronics and programming so that you can quickly begin to build wonderful projects.

If you want to know how to build electronic components driven by a microcontroller, this book is right for you.

Chapter 1
Getting Started with Arduino

Perhaps you have come across the term "Arduino," but like what many people do, you did not bother to find out what it really is. Now, you have some projects that you want to accomplish through Arduino, but you must first learn and understand how Arduino works.

Well, in this chapter I will briefly introduce you to some of the Arduino concepts so that you have a basic knowledge. In the rest of the chapters, we will have a detailed look at everything about Arduino.

Let us first begin by learning what Arduino is.

Arduino is an open-source physical computing setup based on the input/output board together with a platform that executes the processing language.

Arduino can allow one to develop independent devices or link a software stored in the computer. The software can include Flash, MAX, Processing, or VVVV. You can purchase preassembled boards or choose to assemble the boards by hand. If you want to download the open-source

Integrated Development Environment (IDE), you can visit Arduino's website at www.arduino.cc and download it free.

Some of the reasons that make Arduino unique from the rest of the existing platforms in the market include:

- The IDE is simple to use. The IDE is a simple-to-use development platform mostly used by designers and artists.

- It offers a multiplatform environment. This means that regardless of whether you are running on Linux, Windows, or Mac, you will still be able to use Arduino.

- Arduino allows one to program it through a USB cable but not a serial port. This feature is important because many modern computers do not have serial ports anymore.

- Arduino exists as open-source hardware and software. Therefore, if you would like, you can download the circuit diagram as well as purchase all of the components or develop your own without paying the creators.

- Arduino has a large community of active users. Therefore, if you experience a problem, you can ask for help from the community.

- Arduino hardware is at an affordable price. This means you can simply start over by buying a new one when you make critical mistakes.

- The Arduino project was built for an educational setting. It is a fantastic thing for beginners to get things going.

Interaction Design

In the current world, interaction design is essential to help build excellent experiences between objects and humans. It is a great way to dig deep into the development of beautiful as well as mind-blowing experiences between technology and us. The interaction design supports the design process with the help of an interactive process that depends on prototypes. Interaction Design propels design through an interactive process that relies on prototypes for increasing assurance.

This method also belongs to some conventional design types that can be drawn-out to have a prototyping technology, especially with electronics. The exact field of Interaction Design that is concerned with Arduino is called Physical Computing.

Definition of Physical Computing

When it comes to physical computing, we use electronics to develop prototypes of new materials owned by artists and designers.

In other words, it is associated with the design of interactive objects that can exchange communication with human beings. The communication exchange is made possible by sensors and actuators. Both are executed as software installed in the microcontroller.

Traditionally, the use of electronics required the presence of engineers every time. It was difficult for one to build a circuit without an engineer to oversee. These issues inhibited creative people in making the best of their artistic ability. Many of them could not try out different things that they were curious about.

This is because many of these tools were created for engineers and one had to have an in-depth knowledge before they could move on to use. Fortunately, with the rise of microcontrollers, they are among the cheapest and easy to use devices.
The microcontroller has further allowed the construction of better devices.

In fact, Arduino has brought these tools to novices. This will help them learn how to create devices within a short period.

With Arduino, an artist or designer has the opportunity to learn and master the basics of sensors and electronics quickly and develop prototypes with very little capital.

The Arduino Philosophy

When it comes to the concept behind Arduino, it involves the development of designs instead of just speaking about them. It is a continuous search for a faster and more efficient way that you can construct better prototypes.

Classic engineering depends on a strict procedure of moving from point A to point B. However, when it comes to the concept of Arduino, its beauty lies in exploring the unknown land—in this case, going to point C rather than B.

It is fascinating when you choose to be creative with Arduino. You explore the unexpected areas. This helps one discover excellent ideas on how to create great prototypes.

In the next sections, we look at some of the events, philosophies, and pioneers that have motivated the Arduino development.

Prototyping

This is a significant concept in Arduino. It determines how we make things and construct objects that will interact with the rest of the other objects, networks, and people. We strive to discover practical ways to build affordable prototypes.

Most people who are new to electronics have the notion that they are supposed to know the intricate details of building something. Well, while it might look brilliant, it can also be a waste of energy. Why? Your primary interest is to prove that something can work well and then get funds or at least motivate somebody to get funds to support you in building the actual project.

For this reason, we have the opportunistic prototyping. Do not waste time building something from scratch, a process that calls for in-depth knowledge. Instead of going with that option, we can take ready-made devices and apply the effort and hard work invested by great engineers.

Our greatest secret with technology lies in making as many trials as possible. Try out different things on both the hardware and software. You can do this with or without a goal in mind.

Recycling is one of the most exciting approaches in technology. Collecting old devices and toys then remodeling them to build something useful has proven to be an efficient way to get results.

Patching

As an engineer, you will be fascinated by the modularity and the technique to build a complex system by putting together simple devices. Robert Moog illustrates this process well in his analog synthesizer.

Musicians successfully developed sounds by making endless trials of combining several modules using cables. This method resulted in the old synthesizer appearing like the old telephone switch but joined with multiple knobs that were the right platform for tinkering with sound and innovative music.

According to Moog, it was a process between seeing and discovering. In fact, many believe that musicians did not have any clue concerning the function of the many knobs. However, that did not stop them from trying. They improved their style by removing flow interruptions. By reducing disruptions, it made the process seamless.

Circuit bending

One of the fascinating forms of tinkering is circuit breaking. The great thing with circuit benders is the ability to develop the wildest devices through tinkering away with the technology without having a clue of what is happening in the theoretical side.

Keyboard hacks

Up until today, computer keyboards are one of the best ways one can use to interact with the computer. Tinkerers can therefore develop new ideas where they can interact with the software so that they replace the keys with an individual device that can sense the environment.

If you open a computer keyboard, you will find a simple tool inside. At the center, there is a small board. It looks green or brown and contains two sets of contacts that join the two layers between separate keys.

Now, if you remove the circuit and use the wire to connect the two contacts, you will see a letter starting to show up on the screen.

If you have a motion-sensing device and attach it to your keyboard, there shall be a key press whenever a person moves across the PC. Now, connect this to your favorite software, and your computer will be as smart as the urinal.

There is something good with the Junk.

Many people like to throw away or even abandon old technology devices such as old computers, printers, technical equipment, and weird office machines whenever we have a new technology released.

However, these products seem to have a big market, especially for young, poor hackers, and the ones who are just starting out. This market was evident in Ivrea, the place where Arduino was developed. Before you can build an exciting object from these surplus devices, it is essential to know precisely every type of device.

Toy Hacking

Toys are a fantastic source for cheap technology to reuse and hack, as it was evident in the circuit bending practice discussed earlier. With the increasing number of high-tech toys from China, you can come up with quick ideas.

Collaboration

In the Arduino world, collaboration is an important aspect and principle. By engaging with different users in the Arduino forum, you can be sure to get help and learn more about the platform.

The team at Arduino advises people to collaborate at local levels. They further play a role in collaboration through helping them develop user groups in the different cities they visit. With Arduino, you will enjoy the culture of sharing out and helping each other.

Getting Started with the Microcontrollers

We already defined what a microcontroller is. The most important thing is that the microcontroller has a processor and memory as well as some input/output pins that you can control. Most of the time, it is referred to as General Purpose Input Output Pins.

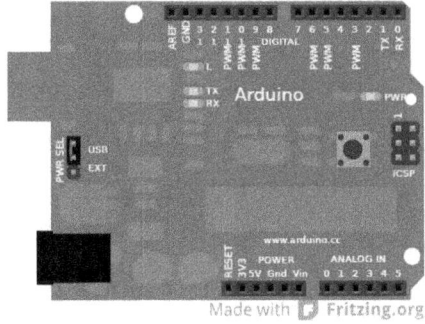

Chapter 2
Hardware and Tools

Arduino is helpful to anyone who would like to program small computers. The tiny machines are called microcontrollers and these improve the level of interactivity.

You encounter a lot of them in your life. You can find it in different devices such as remote controls, toothbrushes, and toys. They perform one primary function, which you rarely see because they do it so well.

The role of the sensors is to listen to the physical world. It converts the energy that you release once you press the buttons, shout, or wave your arms. This energy is converted into electrical signals. Examples of sensors that you can touch with your hands are knobs and buttons, but there are many different types of sensors. Devices that can change electrical energy into physical energy are called actuators.

Microcontrollers talk to actuators and listen to the sensors. It decides on what to perform regarding the type of program written. The electronics and microcontrollers you connect to them are merely a prototype of your projects.

You will have to add your skills so that you can increase the status of the project.

For example, in one project we suggest you create an arrow and connect a motor and enclose it in a box of knobs. In other words, you can build a meter to help people figure out the time that you are available and when you are busy. In a separate example, you will be able to build an hourglass. Arduino is a dynamic component that enhances the usability.

The function of Arduino is to help one build different devices. To make that possible, we did not cover so much on programming and electronics. However, if you are interested in learning much about these fields, you can look for useful guides online. Here we give you some reference sites where you can visit to learn more. One of these places is www.arduino.cc/starterkit

The Components of the Arduino Kit

Breadboard

You can use this type of board for designing electronic circuits. It contains rows of holes that allow one to connect components and wires. Still, there are specific models that require soldering.

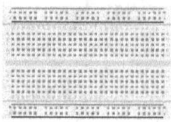

Arduino Uno

This is the microcontroller board that forms the central component of your projects. It is a small type of computer with no means of interacting with you. You will construct the interfaces and circuits and create a means by which the microcontroller can communicate with other devices.

Battery Snap

It connects a 9V battery into the power leads that you can quickly plug into the breadboard.

Capacitors

You must have come across the term *capacitors* if you did a course on electronics. The function of a capacitor is to store and release electrical charge in a circuit. When the voltage in a circuit is more than what is in the capacitor, current flows to the capacitor. However, if the voltage in the circuit is lower, the charge stored in the capacitor flows to the circuit. The capacitor is always connected between the power and ground near the motor to help regulate voltage fluctuations.

DC Motor

The function of the DC motor is to convert electrical energy into mechanical energy. The motor has wire coils that are magnetized when current flows through it. The flow of current creates a magnetic field that attracts and repels magnets. This causes the shaft to rotate. If you change the direction of electricity, the motor will start to spin in the opposite direction.

Diode

The function of a diode is to make sure that electricity flows in one particular direction. It is of great significance if you have a motor in a circuit. Diodes are polarized. This means that the direction in which you position them in a circuit is important.

If you place the diodes in one way, it will permit the current to flow. If you exchange the direction, they block the current. A diode has both the anode and cathode. The anode is connected to the highest energy in the circuit. On the other hand, the cathode is connected to the side with the lowest power. It is easy to identify the cathode because it has a band labeled on one side of the device's body.

Gels

Gels reduce different light wavelength. If you combine gels and photoresistors, it causes the sensor to react to the filtered color.

H-bridge

This circuit helps one regulate the polarity of the voltage used. The H-bridge comes as an integrated circuit, but again, one can build it using some discrete devices.

Jumper wires

Jumper wires are useful when you want to connect devices to each other on the breadboard as well as the Arduino.

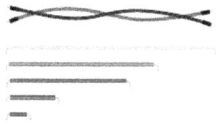

Light Emitting Diodes (LED)

An LED is a type of diode that lights up when an electric current is introduced. Like other types of diodes, electric current only flows in one particular way through these devices. There is a good chance that you are very familiar with these electronic devices. The anode is connected to the power and has a long leg whereas the cathode has a short leg.

Liquid Crystal Display (LCD)

An LCD is a unique type of alphanumeric display that depends on liquid crystals. LCD comes in many shapes, sizes, and styles. The LCD you find in the Arduino kit has 2 rows and 16 characters each.

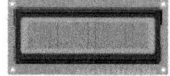

Male header pins

These pins will fit well into the female sockets. Their purpose is to help one connect things easily.

Optocoupler

This helps one connect two circuits that do not have a central power source. Internally, it contains a small LED, which, when illuminated, makes the photoreceptor close the internal switch. When voltage flows through the + pin, it lights up while the internal switch closes.

Piezo

You can use this particular device to produce noise and detect vibrations.

Photoresistor

Also called a light-dependent resistor or photocell, it changes the value of the resistance depending on the intensity of light that strikes its surface.

Potentiometer

It has three pins. Two pins are connected to the end of a fixed resistor while the middle pin moves across the resistor to divide it into two parts. In the potentiometer, when the external sides of the potentiometer are connected to the ground, the middle leg will create a voltage difference when you turn on the knob.

Pushbuttons

These types of switch close a circuit. They can be fitted into the breadboard quickly. They are useful for determining whether a signal is on or off.

Resistors

Resistors reduce electrical energy in a circuit by changing the voltage and the current. The SI unit for resistance is ohms. A resistor has colored stripes to indicate the value.

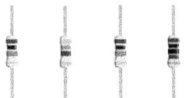

Servomotor

This motor can rotate at 180 degrees. Usually, if you want to control it you have to send an electrical pulse originating from the Arduino. These types of pulses make the motor aware of which position to move.

Temperature sensor

This device varies the output voltage depending on the temperature of the component. The external legs connect to power and the ground. The voltage that passes through the pin changes when it gets warm or cold.

Tilt sensor

This switch will open or close based on the mode of orientation. It has a hollow cylinder with a metal ball inside to complete the circuit when it is tilted correctly.

Transistor

This is a three-legged component that works as an electronic switch. It is a handy device especially if you want to control components that have a high voltage. One pin connects to the ground while the other one connects to the device itself. The last pin is connected to the Arduino. One pin connects to the ground, the other to the device and the last one relates to the Arduino.

USB cable

The USB cable facilitates a connection between the Arduino Uno and your PC to support programming. Furthermore, it supplies power to the Arduino to drive most projects in the kit.

Table of Symbols

These are some of the symbols that you will see throughout this book.

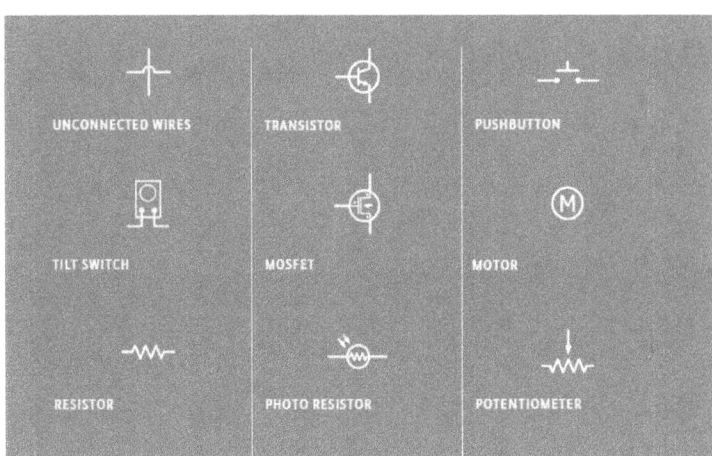

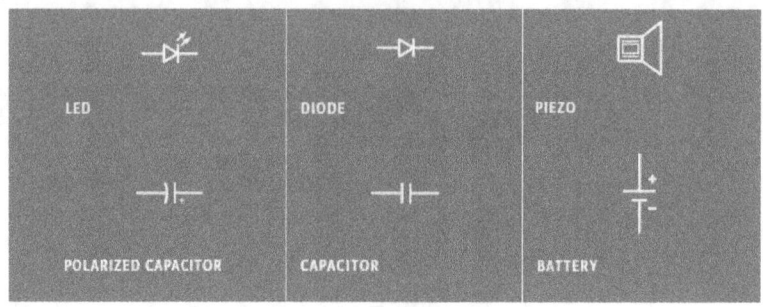

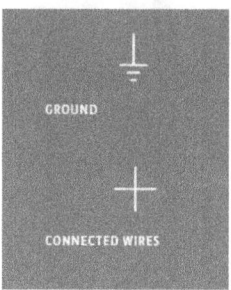

The Board

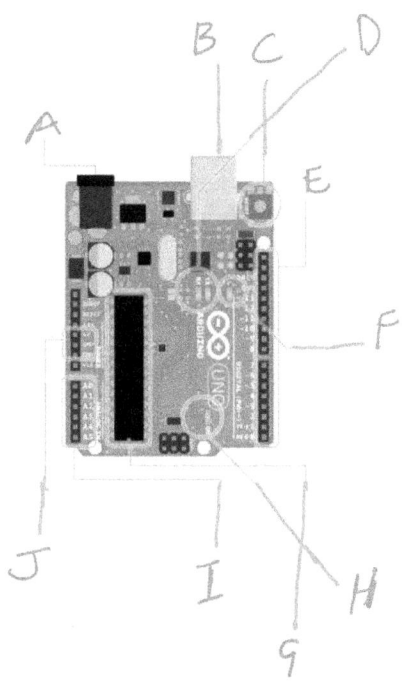

A - Power Connector

This is used to power the Arduino when it is not connected to the USB port for power. It can take voltages in the range of 7-12V.

B - USB Port

Use this to power your Arduino Uno, send sketches to your Arduino, as well as communicate with your Arduino sketch.

C - Reset Button

It resets the ATmega microcontroller.

D - TX and RX LEDs

These LEDs display communication between the Arduino and the computer. You should be ready to see it blink rapidly while you upload the sketch. It is essential when you want to debug.

E - Digital Pins

These pins are applied in the function digitalWrite(), digitalRead(), and analogWrite(). The function analogWrite() works on the pins that have the PWM symbol.

F - Pin 13 LED

This is the only actuator built into your Arduino Uno. It is a great one when it comes to making the first blink sketch. Aside from that, it is handy for debugging.

G - ATmega microcontroller

This is the heart of the Arduino Uno.

H - Power LED

It shows that the Arduino is receiving power. It is essential for debugging.

I - Analog in

Use the following pins with the analogRead().

J - GND and 5V pins

Use the following pins to produce a +5V power to your circuits.

The Arduino starter kit consists of an easy-to-assemble wooden base that makes it easy to work on your projects, no matter whether the projects are from this book or not.

To create it, remove the Wood sheet from the box and follow the instructions below. Make sure that you only use the parts indicated, but do not lose the other pieces. There are specific projects in which you will need to have them.

Well, let's begin:

1. Pick the wood sheet and slowly separate the pieces.

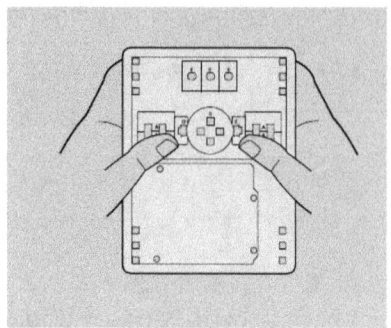

2. Continue until the parts are separated.

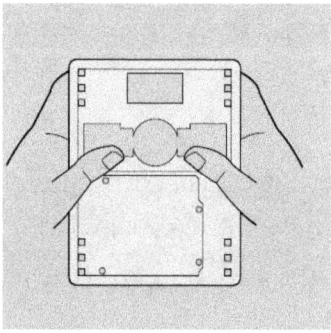

3. Arrange the pieces marked A into the holes in the corners so that you can develop the base feet.

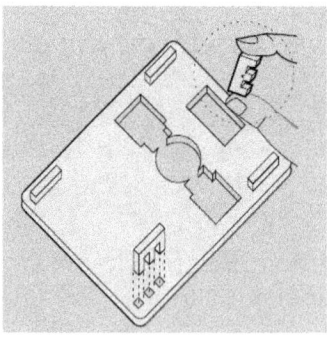

4. Join the Arduino Uno with the base with the help of three screws. Make sure you do not tighten it too much.

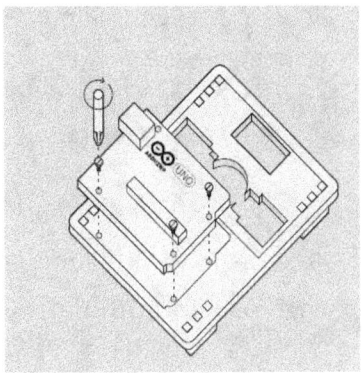

5. Remove the peel from the breadboard carefully.

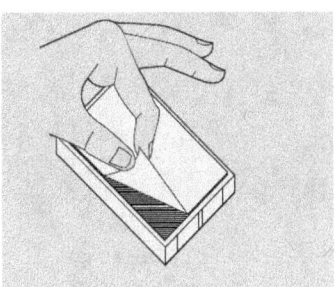

6. Place the breadboard on the wooden sheet close to the Arduino UNO

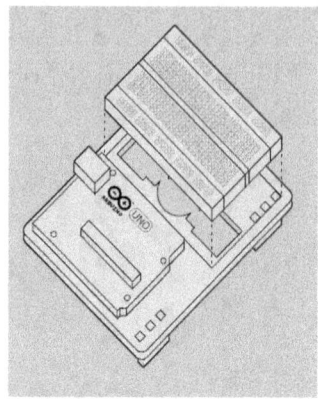

Some of the things that you should have:

- 9v Battery
- Scissors
- Tape and glue
- Small light source
- A box to create holes in
- Old CD or DVD
- Colored paper
- A conductive material such as copper mesh or aluminum foil
- Soldering iron
- Any battery powered component that has at least a pushbutton and a switch
- Basic tools such as a screwdriver

Set up Your Arduino

You don't just start without any order. First, you must establish that your Arduino can communicate or send signals to the computer. This will help one to run the Arduino code.

Windows Installation of the Arduino Package

To ensure that you install the Arduino software, you should download the latest software version that is compatible with your Windows system. To get the newest version, navigate to the download page. Once you are there, you can pick either the Installer (.exe) or the Zip packages. We recommend that you go with the former, which will install everything that you must have for the Arduino Software.

However, with the zip package, you will need to do a manual installation of the drivers. The Zip file is also necessary when you want to have a portable setup. Once the download ends, you can resume the installation. Kindly allow the driver installation process whenever you receive a warning from the operating system.

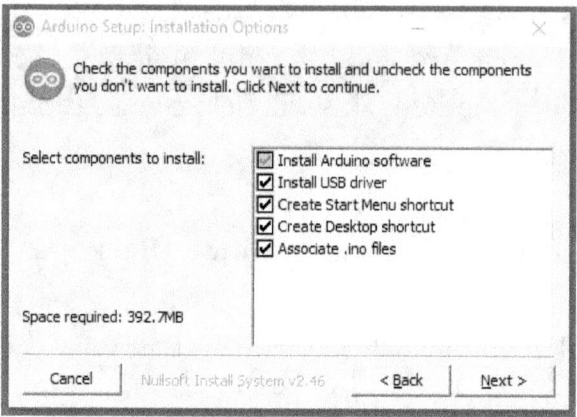

Select the components that you want to install.

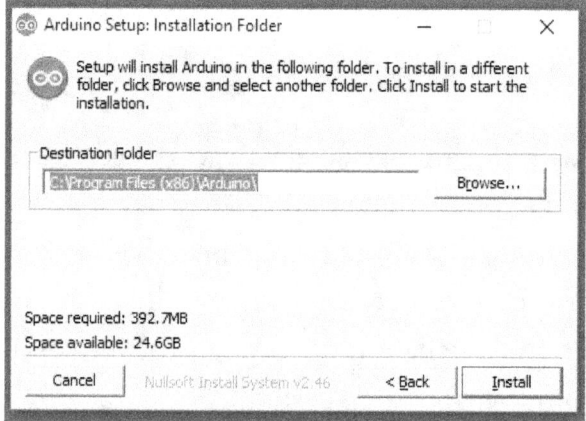

Select the directory to install. It is recommended to stick with the default one.

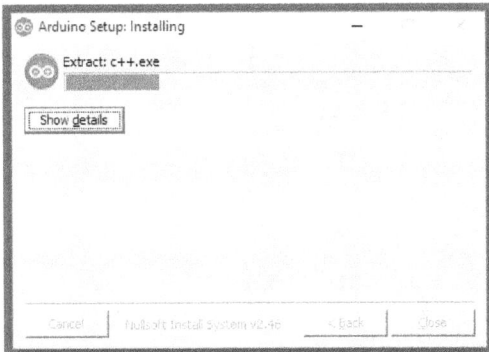

The above process will extract and at the same time install the files to help execute correctly the Arduino Software.

Once you install the IDE correctly, you can navigate to the Getting started Home and pick your board from the available list on the right side of the page.

Mac OS Installation of the Arduino Software

1. Download the IDE and double-click the zip file to open the Arduino application.
2. Transfer the Arduino application into the Applications folder or any other location where you would like to install the software.
3. Next, use a USB cable to connect the board to the PC. The green power labeled PWR should turn on.
4. There is no need to download drivers for the board. A pop-up window may show up to request whether you would like to open the Network Preferences button.

5. The Uno will display the message "not configured," but still it will work correctly. Thus, you can close and exit the System Preferences.

Installation on Ubuntu or Linux set up.

If you want to install the Arduino software package into your Ubuntu OS, first, you must install gcc-avr as well as the avr-libc using a few commands.

Pick the correct JRE if you have more than one JRE installed. Navigate to the main Arduino website and search for the Arduino software manufactured for the Ubuntu. You can proceed to extract and run the software using several commands such as

"tar xzvf Arduino-x.x.x-linux64.tgz cd Arduino-1.0.1. /Arduino"

No matter the type of OS you are using, the instructions above assume that you own an Arduino Uno board. If you decide to purchase a clone, you will get third-party drivers. **Communication with the Arduino**

Once you have installed the IDE and confirmed that you can communicate with the board, the next thing to do is to verify if you can upload the program.

1. Double-click on the Arduino to expand it. In case the IDE opens in a different language that you do not understand, you can make the changes in the application setting. Just navigate to the Language Support, and you will find the details.
2. Switch to the example of the LED blink sketch. A sketch is a term for Arduino programs. You can see it at File>Examples>01.Basics>Blink

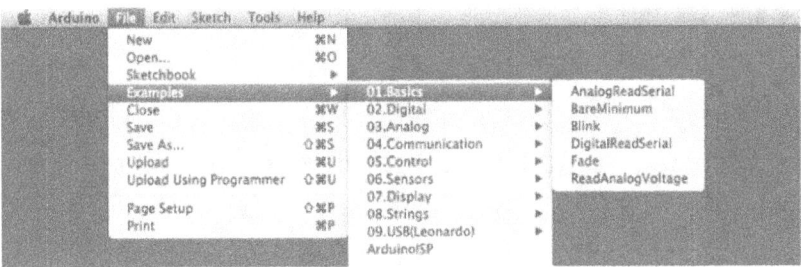

3. You should see a window with some texts popping up. Leave the window, and pick the board below:

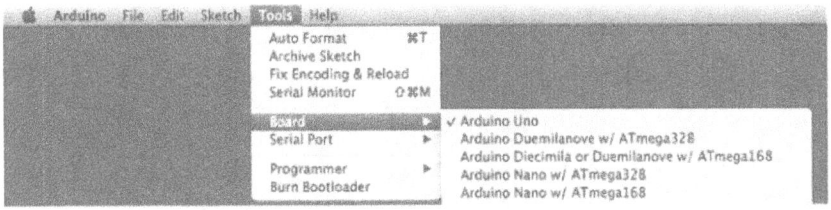

4. Select the serial port which your Arduino is connected to by clicking:

   TOOLS > SERIAL PORT

If you are using Windows, look for the COM that has the largest number. Do not be afraid to select the wrong COM. If the first one fails to work, you should try the next one. For verification purposes, disconnect the Arduino board and re-open the menu. Re-connect the board and select the serial port.

For those who are using the Mac, they should navigate to the /dev/tty.usbmodem in it. Generally, you will see two of these, so select either of it.

1. If you want to upload the Blink sketch to the Arduino, navigate to the top left corner and click UPLOAD.

2. You should see a status bar that shows the progress of the upload at the lower side of the IDE. The lights should be labeled RX and TX on the Arduino board. Once the process of uploading ends, a message will pop up to remind you of the end of the upload process.

3. After a few seconds when the upload is over, you will see a yellow LED with the letter L starting to blink. Congratulations! You have successfully managed to program the Arduino so that it flashes its onboard LED.

We have individual cases where you purchase the Arduino that is configured with the Blink sketch. In this situation, you will not know whether you are in control. Update the delay time to 100, and upload again the Blink sketch. This time around, the LED should blink faster.

Understand your tools

You will build a simple circuit that consists of switches, an LED, and a Resistor. In the end, you should understand some basic electrical theory, the way a breadboard operates as well as devices in parallel and series.

Electricity is similar to other energy types such as gravity, heat, and light. Electrical energy flows through conductors such as a wire. It is possible to transform electrical energy into other energy types to perform specific tasks such as turn a light on or produce noise in a speaker.

Devices that do these such as light bulbs or speakers are called electrical transducers. A transducer converts other forms of energy into electrical energy. Devices that convert other energy forms into electrical energy are called sensors, and devices that convert electrical energy into other energy forms are called actuators. You will learn how you can create a circuit to transfer electricity through components of different types. Circuits are closed wire loops that have a power source and something that uses the energy called a load.

Electricity flows in a circuit from a point of higher potential energy to an end of lower potential energy. The ground is the point with the least energy in a circuit. In the circuit that you create, electricity will flow in one particular direction.

This type of circuit is called a direct current or DC. For the Alternating Current, the voltage changes the route in an interval of between 50 to 60 second.

You need to know about several terms when working with electrical circuits. Current refers to the size of electrical charge that flows past a certain point in the circuit. The units for current is amperes or the A symbol. Voltage is the difference between a given point in a circuit and the other one. Then we have resistance measured in ohms. Resistance refers to the level in which a device prevents the flow of electrical energy.

A great way to imagine this is by thinking of a rock that is moving down a cliff. If the cliff is very high, then the rocks will have a lot of energy when they fall at the bottom.

In this case, the voltage in the circuit is equal to the height of the cliff. When the voltage is high at the source, the more power you are going to use. The higher the voltage at the source, the more power you will have to use. The more number of rocks, the higher the energy is transferred down the cliff.

The amount of stones is equal to the current in the circuit. Rocks roll through the bushes on the side of the cliff, and in the process, it loses some energy.

The energy lost is used to demolish the forests. The forests are equivalent to the resistors; it produces a given opposition to the electrical flow and changes it to other forms of energy. Below is a diagram to illustrate:

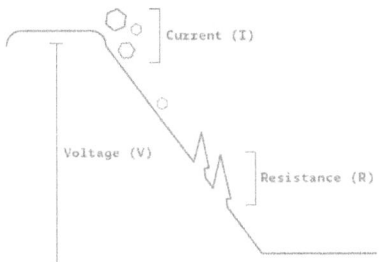

There has to be a complete path that begins from the energy source and goes to ground to make a circuit. If there is no path for the energy to flow, the circuit will not work.

The components in an electrical circuit consume all the electrical energy. Every element will convert part of the power into a different type of energy. In each circuit, every voltage is transformed into a different form of energy.

The path for the current at any given point in a circuit is similar both in and out.
Electrical current will search for the path with the minimum resistance to the ground. With the existence of two possible routes, most of the electrical current will flow down the path that has the least resistance.

Now, if power is connected to the ground without any resistance, chances are that you will short circuit. If a short circuit takes place, the source of power and the wires will convert electrical energy into heat and light.

Typically, it appears as an explosion. If you have ever tried to short-circuit a battery, then you know how dangerous short-circuiting can be.

The key place to create circuits is the breadboard. The one in the kit is solderless. It is solderless because there is no need to solder anything. You can look at the horizontal and vertical rows of the breadboard below:

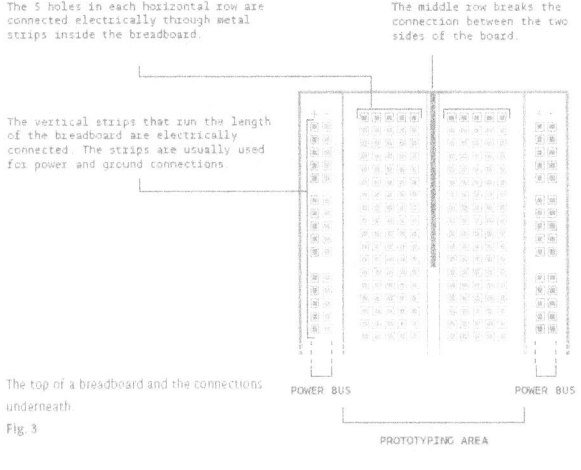

Fig. 3

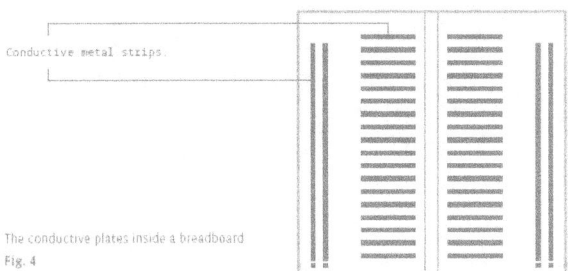

Fig. 4

Circuit Drawings

In this project, you will see two types of circuits: the one in breadboard view that resembles the device inside your kit. The other one is a schematic view that displays the relationship between components in a circuit. Schematics will never show the location of elements to each other, but it reveals how it is connected.

Breadboard view

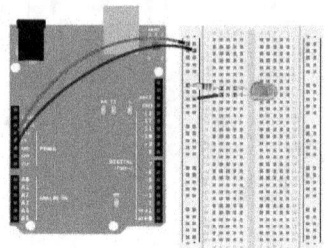

Schematic view

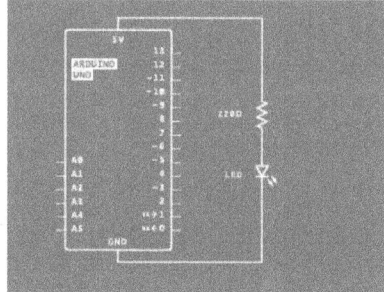

1. Connect the breadboard to the Arduino's 5V and the ground connections. Install two LEDs, and a green LED into the breadboard. Connect the cathode for each LED to the ground.

2. Place the switch on the breadboard. Connect one to the power and the other one to the digital pin of the Arduino. Also, get a 10k-ohm resistor and connect it to the switch pin of the Arduino. The pull-down resistor joins the pin to the ground when the switch

opens, and this indicates LOW when the voltage passes through the switch.

You can decide to cover the breadboard using a template from the Arduino kit. Alternatively, add some creative designs to build your launch system.

Each Arduino program contains two significant functions. The functions belong to the computer program that executes individual commands. Functions have special names and are called when required to accomplish a given task. The most critical functions in the Arduino program are the setup() and the loop(). To use these functions, you must declare them first. This means that you will have to explain to the Arduino its function.

In this program, you are going to define a variable before moving to the major parts of the program. Variables refer to names you assign to places in the memory of the Arduino to help monitor what is going on. These values are bound to change at any time depending on the instructions of the program.

It is essential to have the variable names describe a lot the value it contains. For instance, a variable switchState describes the kind of value it stores. This could be the status of the switch while a variable like "y" does not reveal much about its value.

For one to create a variable, you must first declare its type. Variables can be of type *int*, which contains a whole number. When declaring a variable, you have to assign it an initial value. After every variable declaration, you have to put a semicolon at the end.

The setup() function runs only once after you power the Arduino. This is the correct time to configure the pins by using the function pinMode(). The pins that are connected to the LEDs contain the OUTPUTs while the switch pin is the INPUT.

The program contains the loop() function which frequently runs when setup() completes. The loop() helps one to verify the voltage on the inputs, and switch the outputs on and off. To validate the amount of voltage in a digital input, use the digitalRead() function.

If you want to know the type of pin to look, you have to supply an argument to the digitalRead() function. Arguments are information that you pass to the functions. The info should help the function perform its job well. For instance, the function digitalRead() requires an argument. The type of pin that you should go to check is operated by digitalRead().

In the code, we used the word if to find the status of something. In programming, an if statement will help one compare two things and determine the accuracy of the comparison. Then it will complete the actions send to it. When you want to compare two elements in programming, the == signs are applied. If you use one, then you will be assigning a value rather than making comparisons.

The digitalWrite() function will allow one to send 0V or 5V to a specific output pin. The digitalWrite() accepts two parameters: the type of pin to control, and the value to set on the pin. The value can be HIGH or LOW.

You have communicated to the Arduino on what to do whenever the switch is open. Therefore, it is time to define what will happen when you close the switch. The if() statement contains an optional else part which will allow something to take place when the initial state fails

To find out whether the red LEDs blink when you press the button, you must switch the lights off and on in the else section. Once you change the LEDs to a specific state, wait for some seconds before you change the Arduino back.

However, if you do not wait for some seconds, the lights will rapidly move back and forth such that will look like it is dim but not off. One reason for this is that the Arduino will run through the loop() more than a thousand times per second and the LED will turn on and off faster than we thought. The delay() function prevents the Arduino from running any operation for a short interval.

```
void setup(){
}

void loop(){
}
```

{ Curly brackets }
Any code you write inside the curly brackets will be executed when the function is called.

```
1  int switchState = 0;

2  void setup(){
3      pinMode(3,OUTPUT);
4      pinMode(4,OUTPUT);
5      pinMode(5,OUTPUT);
6      pinMode(2,INPUT);
7  }

8  void loop(){
9      switchState = digitalRead(2);
10     // this is a comment
```

Case sensitivity
Pay attention to the case sensitivity in your code. For example, pinMode is the name of a command, but pinmode will produce an error.

Comments
If you ever want to include natural language in your program, you can leave a comment. Comments are notes you leave for yourself that the computer ignores. To write a comment, add two slashes //. The computer will ignore anything on the line after those slashes.

After you are complete programming your Arduino, you should see a green light. If you touch a button on the switch, the red light will begin to flash while the green light turns off.

Attempt to change the time of the delay() function; pay attention to how the lights will behave and the way the system response will vary based on the speed of the flashing. If the delay() function is called in the program, it will stop the rest of the functionality.

There will be no sensor readings until when the time ends. Even though days are useful, when you create your projects, it is critical to ensure that they do not disrupt your interface.

In this chapter, you have learned how you can set up your first Arduino program that changes the behavior of some LEDs based on your switch.

You have learned how to use an if…else statement, variables, and some functions.

Chapter 3
Turn your Arduino into a Machine

While switches and buttons are an excellent thing, still there is a lot to do than turn it on and off. Although Arduino is a digital device, it can receive information from an analog sensor so that it can measure light, temperature and so on. To create this, you use the built Analog-to-Digital Converter of the Arduino.

You will use a temperature sensor to determine the warmth of your skin. This device releases a changing voltage based on the temperature it detects. It comes with three pins: the first pin connects to the ground while the other connects to the power. The third pin transfers the voltage of the variable into the Arduino.

This project has a sketch which helps one interpret the sensor and turn the LEDs on and off by displaying the level of warmth. Temperature sensors are of different types. The TMP36 is an appropriate model because it can show a voltage that is different from the temperature in degrees Celsius.

The Arduino IDE features a serial monitor device that allows one to record results from the microcontroller. Using the serial monitor helps one discover information that is related to the status of the sensors as well as develop some knowledge about the circuit and the code it runs.

Create the Circuit

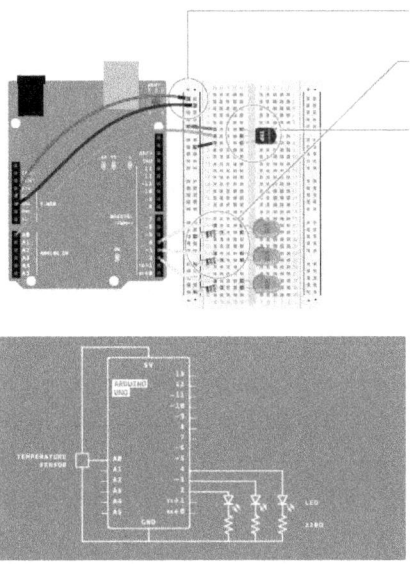

In this project, you will learn how to find the ambient temperature of a room. Now, you are doing it manually, but you can achieve that by calibration. You can use the button to define the reference temperature, or let the Arduino pick a sample before a loop() starts and have it as the point of reference.

1. First, connect your breadboard to the ground.
2. Connect the cathode of every LED you have to the ground using a resistor. Join the anodes of the LEDs to pins 2 using 4. These are the project indicators.

3. Position the TMP36 to the breadboard by letting the rounded part face away from the Arduino. Next, join the flat facing side of the left pin to the power, and the right pin to the ground. Connect the central pin to the AO on the Arduino.

Build an interface for the sensor to help people use it. You can use a paper cutout that resembles a hand of a good indicator. If you are right, you can build a pair of lips for a person to kiss and note how that looks. You might also want to mark the LEDs so that it can reveal some meanings.

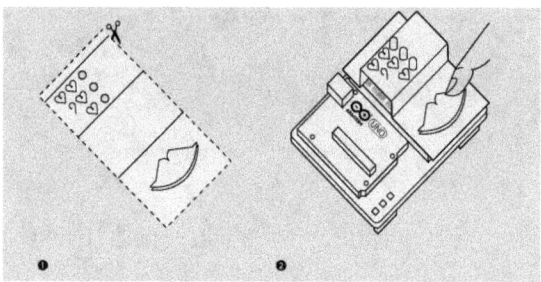

1. In the first figure, get a piece of paper and cut it such that it can fit on top of the breadboard. Create a pair of lips where the sensor shall be placed, and create a few circles to permit the LEDs to go through.

2. Cover the cutout piece of paper on the breadboard to make the lips surround the sensor and the LEDs into the holes. Press the lips to see how it feels.

Let's examine the Code.

Useful constants

Constants allow someone to give unique names to things in the program. This is similar to variables except that they cannot change. Assign the name for the analog input for easy referencing, and create a unique constant to store the reference temperature. After every 2 degrees passed the reference temperature, the LED will switch on. Temperature is written and stored in a floating-point number. A floating-point number is one that has a decimal point.

Initialization of the serial port

In the setup, you will interact with a new command called Serialbegin(). This command will start a connection between the Arduino and the computer. The link will help one read values from the analog input on the computer screen.
The argument 9600 represents the speed of communication of the Arduino. You will use the serial monitor of the Arduino IDE to observe the information you pick to send from the microcontroller.

Initialize the digital pin and switch it off

The next thing is the for() loop sets a few pins as the output. These pins were previously connected to the LEDs. Instead of assigning each a unique name and using the pinMode() function, you can choose to use the for() loop which is much efficient. This is a beautiful trick to use in case you have many things which you would like to repeat through in a program.

Reading the sensor temperature

While in the loop(), use the variable called sensorVal to hold the sensor reading. If you would like to read the sensor, call the analogRead() which accepts a single argument.

Transfer the sensor values to the PC

The Serial.print() function transfers data from the Arduino to the PC. You can check this information in the serial monitor. If you assign the Serial.print() a parameter in the quotation marks, it displays the text typed. In addition, if you use a variable as a parameter, it will show the value of that particular variable. Below is the code for the program:

```
1  const int sensorPin = A0;
2  const float baselineTemp = 20.0;




3  void setup(){
4    Serial.begin(9600); // open a serial port




5    for(int pinNumber = 2; pinNumber<5; pinNumber++){    for() loop tutorial
6      pinMode(pinNumber,OUTPUT);                          arduino.cc/for
7      digitalWrite(pinNumber, LOW);
8    }
9  }

10 void loop(){
11   int sensorVal = analogRead(sensorPin);

12   Serial.print("Sensor Value: ");
13   Serial.print(sensorVal);
```

Convert the sensor reading into a voltage

With some knowledge of math, you can determine the right pin voltage. The voltage can range from 0 to 5 volts and has some fractions. You will have to declare a float variable to store it there.

Changing voltage to temperature before uploading to the PC

The sensor's datasheet has information similar to the output voltage. Datasheets are like electronic manuals. They are created by engineers to be used by other engineers. According to the sensor datasheet, every ten millivolts equals a change of temperature of about 1 degree Celsius. Furthermore, the sensor can read a temperature that is below 0 degrees. Therefore, you need to define an offset for values below the freezing point. If you are to minus 0.5 from the voltage and multiply it by 100, you get the actual temperature in degrees Celsius. Create a floating-point variable and store the new number.

Now you have the initial temperature with you, and you can print that in the serial monitor. Given that the variable that stores temperature is the last thing printed in the loop, you have to use a separate function such as a Serial.println(). This command will help one build a new line once it sends a value. This will simplify everything when it is printed out.

Turn off the LEDs for low temperature

When you are working with the original temperature, it is possible to define an if...else statement to turn on the LED. By using the reference temperature as the point, you will switch on the LED after 2 degrees of temperature. You will scan for a range of values as you look through the temperature scale. Below is the next part of the program.

```
14    // convert the ADC reading to voltage
15    float voltage = (sensorVal/1024.0) * 5.0;

16    Serial.print(", Volts: ");
17    Serial.print(voltage);

18    Serial.print(", degrees C: ");
19    // convert the voltage to temperature in degrees
20    float temperature = (voltage - .5) * 100;
21    Serial.println(temperature);

22    if(temperature < baselineTemp){
23        digitalWrite(2, LOW);
24        digitalWrite(3, LOW);
25        digitalWrite(4, LOW);
```

Starter Kit datasheets
arduino.cc/kitdatasheets

Turn on the LED to create a low temperature

The && operator stands for "and" in the logical sense. It allows one to check for multiple conditions.

To create a medium temperature, turn on the two LEDs

61

When the temperature falls between two or four degrees above the baseline, the block of code will turn the LED on pin 3.

Switch on the LEDs for a higher temperature

The Analog-to-Digital Converter allows one to do read fast. Therefore, you have to create some delays at the end of the loop(). If you constantly read the values, it will appear erratic.

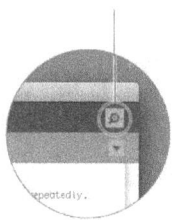

Once you transfer code to the Arduino, you can select the monitor icon shown in the figure above. A stream of values formatted in this manner:

Sensor: 200, Volts: 70, degrees C: 17

will show up.

Rest your fingers close to the sensor and observe carefully on what will happen to the serial monitor. Note down the temperature when the sensor is in the open air.

Close the serial monitor and modify the baseline temp constant to the value you noted. Re-upload the code again, and attempt to touch the sensors with your fingers. When the temperature goes up, the LEDs start to turn on one by one.

Chapter 4
C Language Basics and Functions

When you create an Arduino program, it is essential to have some knowledge about the working of computer systems. Even though C programming is the language that is close to the machines, how certain things are done when the program runs will become clear.

A primary system consists of the control device referred to as the CPU or microcontroller. There are a few differences when it comes to some of these. We shall dig deep into this later. Just to mention, microcontrollers may not be that powerful compared to the standard microprocessor. However, it still contains input, output ports, as well as hardware functions.

Microprocessors are connected to the external Memory. Generally, microcontrollers contain a sufficient amount of onboard memory. However, it should be noted that we are not referring to the large sizes; it is possible for a microcontroller to have only a few hundred bytes or so of memory for the simple applications. Don't forget that a memory byte has 8 bits, and each bit can either be true or false, high or low and I/O.

When it comes to the relation between the processor and the functional data stored in the memory, data must be kept in the processor's register. The register is the only place where we can have logical mathematical operations carried out. For example, if you would like to carry out an addition of two variables, the value of the variables has to be moved over to the register.

Memory Maps

Each memory byte in the computer system has a connected address. Now, if we do not have the address, the processor will not have a means to identify a particular memory. In general, the memory address begins from 0 as it increases. Even though we have specific addresses with a private or unique system, a particular address may not point to the input and output port of external communication.

Most of the time, you will find it necessary to map-out the memory. This is merely a massive array of memory slots. We have people who develop a memory map and have the address with the least value positioned at the top while others who draw a memory map and assign the least address at the bottom. Each address points to a place where it can have the byte stored.

However, the C compiler will complete this. For instance, if we declare the char variable as −X. It can be located at address 2, so if we print the value, there would be no need to select the value at the address 2. We will instead write, "select the X value" where the compiler produces code to ensure that it works correctly to the right address. Using this level of abstraction simplifies the whole process.

However, since most variables carry a specific amount that is higher than one byte, we might have to collect these addresses to hold only a single value. For example, if we pick a short int, then it will require us to have two bytes.

Now, if the following address starts at four, there is a need to use the address 5. When we choose to access this particular variable, the compiler will automatically build the code and make use of all the addresses since it is aware of the presence of the short int.

Stacks

Most programmers prefer to use temporary storages for the variables. What this means is that there are variables that are used for a short period then they are discarded. Therefore, it will not be right if we move on and allocate a permanent space for this particular variable.

Ordinarily, an application is made up of two parts: the code and the data. The data part is permanent since these two parts cannot consume the whole memory; the remaining memory is used temporarily for storage via the stack. It starts at the opposite end of the memory map; the stack increases towards the data part as well as the code.

It is similar to the stack of trays. The first tray on the stack will be the last to be pulled off. Any time temporary variables are needed, this part of the memory is used. Given that many items are required, most of the memory will be used up. When the code ends, the temporary variables declared are no longer useful, and therefore the stack shrinks.

The Basics of C-language

C language is designed for professional developers who want to accomplish many things with less code. C is a compiler language. This shows that once we have written the program, we must transfer it into the compiler that will begin to change the C language instructions into a machine code that the microcontroller can manage.

As you can see, this is an additional step to take, but it will result in a better program compared to the interpreter. After this, the interpreter will convert the code from the machine language.

It is crucial for the machine to have an interpreter. You can look at it as a compiler that translates it once instead of line-by-line.

However, C is not the same as in other languages. It is a free-flow language. We have the statements, functions, and variables. Variables, as we have already seen, are objects that can store things. It can be a floating-point number or other types of variables. Statements have assignments, operations and so on. Functions have statements and can call other functions.

How to name variables, and declare

The naming of variables in C is quite easy. Names of variables carry with them numerals, underscores, and letters. You can as well combine the upper and lower case. However, the length cannot go past the 31 characters.

However, the actual limit depends on the C compiler. In addition, variables cannot contain reserved keywords or unique characters like a semicolon, comma, and other special characters. Therefore, valid names can be names such as resistor8, volt5, and we_are_variables.

C language has different variable types. Some of them consist of floating-point numbers and real numbers in two forms. First, there is the 32-bit float, and then the double. We also have a few types of integers that consist of char, 16 bits, short int, and 32 bits long int. Though char is only an 8 bit, still has a 2 to the 8th combinations – or even 256 separate values that you will find it perfect for a single ASCII character.

Similar to other languages, the C language has arrays and compound data types.

When it comes to variables in the C language, it is vital for the variables to be declared before they are used. Variables cannot just be created instantly like the way it happens in the Python language. Variable declarations are made up of the variable type and variable name. You can also include an initial value for the variable during the declaration, but that is optional. Multiple variable declarations are still allowed in C language. For instance:

Float c = 1.2;
Char c;
Unsigned char x;

You should underline that every type of variable declaration ends with a semi-colon. Like many other programming languages such as Java, the semi-colon indicates the end of that statement.

Functions

Functions have a similar naming rule like variables. All functions have a similar syntax that resembles:

Return_value function_name (function argument list)

{
Statement(s)
}

You can borrow the concept of mathematical functions where you assign it some value(s) as well as allocate back some values. An example is a calculator that contains the cosine function. It is possible to assign an angle to it, and this will return a specific value. Functions may contain separate arguments in the C language. Furthermore, it is possible for a C function to return values. A void function is one that does not require a value or return a value. A *void* function will look like this:

void function_name (void)
{
//necessary statements come here
}

This might look like a lot of work, but the data types in the C language make sense. What this means is that if you choose to use a wrong kind of variable in a function, or even an incorrect number of variables, you will receive a warning.

Therefore, if you have a float function and attempt to send it an integer variable, the compiler will send you a warning. Every program must have a start and an end. In the C language, all programs begin at the main function. You can look at the program below:

```
void main (void)
{
float y = 3.0;
float f = 2.0;
float t;
t=y*t
}
```

There exists one main() function. It accepts no variables and returns nothing.

Libraries

The previous example is limited because it is hard to see the result. Therefore, you will need some methods when you want to display the results on your computer screen. To achieve this, it will depend on the libraries and systems functions. Countless libraries have the most advanced C systems. In essence, somebody has just tested, compiled and wrote a collection of functions. What you need to do is to link the functions into the program. Linking helps integrate the code and any existing library into an entire program. To display the general data and the input data, we use the standard IO and the stdio.

The stdio library has a function called printf().

```c
// Our third program, this is an example of a single line comment

#include <stdio.h>

void main( void )
{
     printf("Hello world.\n");
}
```

The above program will display the words "Hello World" on the computer screen. It will further insert a new line after the backslash-n combo. The \n refers to the addition of a new line. If we failed to include the #include directive, the compiler will not understand anything concerning the printf(), and it would show up an error when we attempt to use it. Well, what about the header file?

The header file has a lot of different function prototypes. These prototypes can be seen as templates but if you want, you can build your own.

To make use of it, you are required to have the correct include statement written into the code, and it will be better set if you remember to include the linker library code. This will not only save time but also allow you to reuse the code.

Simple Math

C has certain basic math operators just like other languages. Some of them include the -, +, / and the multiple.

Parentheses help divide the elements and power hierarchy operations. The C language has the % operator that represents modulo. The modulo is an operation which carries the remainders of a division. For example, 8 modulo 18 would, of course, be 2. The division will behave separately both to integers and floats which have no remainder.

In other words, integer 5 divided by 2 is 2, and not 2.5. Within the C language, there is a sequence of bit manipulators useful for situations such as this. For complex math operators, you will have to go deep into the math library. Some of the examples are log (10), tan(), cos() and sin().

However, you should not attempt to use the ^ operator because it has a separate meaning in the C language. Well, do you still remember what we said earlier about the use of libraries? Placing certain functions such as sin() into your code forces the compiler to define the prototypes along with other related information. Therefore, during the start of the program, it will be necessary to include the following line:

#include <math.h>

C Language Input and Output

We know that the prinf() function displays the information on the screen. The printf() is an extensive and complex function which has a lot of variants and format specifiers. The format specifiers comprise of the % stuff applied as the placeholders for the values. Some of the examples include:

%f	float
%lf	double (long float)
%e	float using exponent notation
%g	float using shorter of e or f style
%d	decimal integer
%ld	decimal long integer
%x	hexadecimal (hex or base 16) integer
%o	octal (base 8) integer
%u	unsigned integer
%c	single character
%s	character string

Take for instance, if we want to show the value of a variable in the decimal form. We could have done it this way:

printf ("The value is %d, in hex %x, and in octal is &o.\n", value1, value1, value1);

You should see the way we have labeled the variables. This is critical because if you make a mistake and display a value that has no label, it will be impossible to tell whether it is a hex or decimal. For instance, if you see a number like 22, how will you tell that the number is a decimal or hex for that matter? It is impossible to know.

Besides indicating the label, you can print it with a field width. For instance, %6d is equivalent to writing the integer in the decimal with a minimum space of 6. Similarly, %6.2f implies that you print the floating-point value with a minimum of 6 spaces. The .2 part is an exact specifier, and in the following example, it shows two digits after the decimal point. You can then see how powerful and flexible this function looks.

The input function for the C language is the scanf(). This resembles the Python's input statement. Even though it is possible to request different values at once, it is the best.

It comes with similar specifiers like the printf(). So, there is a point which needs to be understood, and the scanf() function will require you to specify the location where the value is kept in the computer memory. This shows that just writing the name of the variable is not enough. You need to describe it in detail.

C has the & operator which means the address of. For example, when you want to select a specific integer variable from a user and store it snugly inside the voltage variable. Here is the code fragment for you to look at:

printf ("Kindly type in the voltage");
scanf ("%d," &voltage);

Bitwise operators

There are certain times when you would want to carry out bitwise operations instead of the ordinary math.

For instance, if your goal is to AND two variables bit-by-bit, bitwise operations would be the best to use in writing the code to program the microcontrollers, testing and clearing specific bits in the control registers.

C consists of different bitwise operators. Some of them include AND, XOR, Shift Left, One's Complement and Shift Right.

Chapter 5
Advanced Inputs, Outputs, and Sensors

So far, you know some of the basic operations that you can complete in the Arduino, for instance, to regulate the digital output and digital input.

Other On and Off Sensors

Besides the push button, other primary sensors work on the same principle.

Switches

The same way we have a pushbutton, a switch does not change its state whenever it is released.

Thermostats

This switch opens whenever the temperature exceeds an absolute limit.

Magnetic switches

This contains two contacts that join when they come close to a magnet. They are applied in the burglar alarms to determine when a window opens.

Carpet Switches

They refer to the mats that help you monitor the presence of a human being.

Tilt Switches

This primary electronic component has two contacts and a small metal ball. An example of a tilt switch is the tilt sensor. If the sensor is upright, the ball in the sensor completes the two contacts.

This seems to work as if one has pressed the push button. When the sensor is tilted a bit, the ball will move, and the contact opens. This is similar to the way one has removed a push button.

If you apply this type of component, you can successfully implement the gestural interfaces that react any time an object is shaken. You need to experiment by finding all kinds of devices that have two contacts similar to the thermostat.

Regulate light with PWM

Up to now, you have learned many things; we believe you can use that knowledge to build an interactive lamp.

One disadvantage of the blinking LED is that it only allows you to switch the light on and off. You cannot do any other thing with it. To solve such problems, we can apply a simple trick that makes it possible for a continuous vision.

We can take an example of a blinking LED. If we change the numbers in the delay function, the LED will stop blinking. In fact, you will realize that the LED begins to look dim at 50% of the standard brightness.

You have to change the numbers to make the LED remain on within a quarter of the time when it is off.
This method is referred to as the pulse width modulation.

It is an interesting approach of saying if you blink the LED very fast, you will not see it blink anymore. However, you can modify its brightness by updating the time that it is on and the time when it is off. This method still works with devices such as the LED. For instance, it is possible to change the motor speed in the same manner.

During the experiment, you will realize that when you make the LED to blink, and then introduce specific delays in the code, it becomes inconvenient. This is because when you transfer data via the serial port, the LED changes when it waits for one to finish reading the sensor.

Luckily, the type of processor used in the Arduino board has a hardware piece that can blink three LEDs if your sketch does something different. The hardware is implemented in the pins numbered 9, 10, and 11. They can all be controlled by the analogWrite() instruction.

Use Light Sensor instead of a Pushbutton

Now, we want to attempt an exciting experiment. Pick a light sensor. In dark settings, the resistance of light-dependent resistor goes high. Whenever you shine the light towards it, the resistance will drop very fast and become a good electricity conductor. In other words, it is something similar to a light activated switch.

1. Create the circuit with the help of a Pushbutton to regulate the LED.
2. Next, insert the LDR into the breadboard instead of the pushbutton. You will notice that anytime you cover the LDR with your hands, the LED remains off.
3. When you uncover the LDR, it turns on. Congratulations, you have successfully built the first real sensor powered LED.

The Analogue Input

So far, we know that the Arduino can help one to check whether the voltage is flowing into the pins. Furthermore, it can measure the voltage using the digitalRead() function. The light sensor that we have used can show us if there is light. Besides that, it can display the amount of light present.

This creates the difference between on/off sensor. Its value will always change so that it can read this particular type of sensor.

When you look at the lower part of the Arduino board, you will see six pins labeled Analog in. These are unique types of pins which can show not only the existence of voltage but also the voltage value. Now, if you apply the analogRead() function, you read the voltage directed to either of the pins.

This powerful function returns the number in the range from 0 to 1023. This represents the voltage between 0-5 volts. For instance, if we have a voltage of 2.5V directed to a pin 0. The analogRead (0) will display 512.

If you create a circuit like the one shown below and run the code in the following section. You shall see the onboard LED blinking an LED at a speed that is determined by the levels of light that hit the sensor.

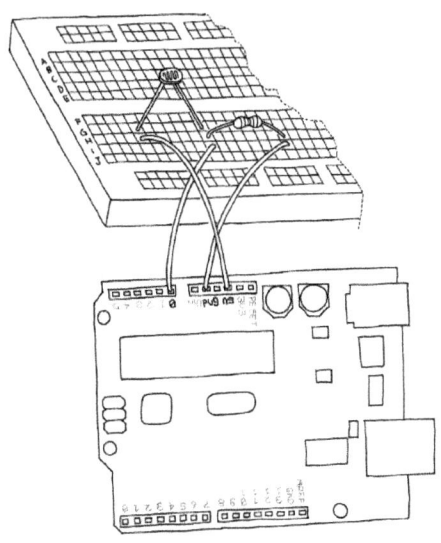

```
// Example 06A: Blink LED at a rate specified by the
// value of the analogue input

#define LED  13 // the pin for the LED

int val = 0;   // variable used to store the value
               // coming from the sensor
void setup() {
  pinMode(LED, OUTPUT); // LED is as an OUTPUT

  // Note: Analogue pins are
  // automatically set as inputs
}

void loop() {

  val = analogRead(0); // read the value from
                       // the sensor

  digitalWrite(13, HIGH); // turn the LED on

  delay(val); // stop the program for
              // some time

  digitalWrite(13, LOW); // turn the LED off

  delay(val); // stop the program for
              // some time
}
```

Other Analogue Sensors

If we use the previous circuit, we can connect multiple resistive sensors that can work in the same way. For example, you may connect the thermistor—a little device whose resistance will change in temperature.

In the circuit, we have discussed the way changes in the resistance change the voltage that is measured by the Arduino.

If you choose to use a thermistor, it is essential to understand that we do not have a direct relationship between the value that you read and the initial temperature measured. If you want a correct reading, you need to look at the numbers that appear out of the analog pin when you measure the real thermometer. You could still place the same numbers on the table side by side and build a method where you can mark the analog results to the actual temperature of the world.

Up until now, the LED has been used as an output device. However, you might want to know how to read the initial values accessed by the Arduino from the sensor.

It is difficult to make the board blink when the values are in the Morse code. However, there is an easy way where humans can read the values. In this example, it is possible to let the Arduino communicate via a serial port.

Serial Communication

We once said somewhere in the Book that the Arduino has a USB connection that is connected to the IDE to send the code to the processor. The best thing is that the connection can be established using the sketches to post the code to the computer.

The object consists of the code we need to transfer and receive the data. Enter the code below into a new sketch.

```
// Example 07: Send to the computer the values read from
// analogue input 0
// Make sure you click on "Serial Monitor"
// after you upload

#define SENSOR 0  // select the input pin for the
                  // sensor resistor

int val = 0;  // variable to store the value coming
              // from the sensor

void setup() {
    Serial.begin(9600);  // open the serial port to send
                         // data back to the computer at
                         // 9600 bits per second
}

void loop() {
    val = analogRead(SENSOR);  // read the value from
                               // the sensor

    Serial.println(val);  // print the value to
                          // the serial port

    delay(100);  // wait 100ms between
                 // each send
}
```

Once you have this code in the Arduino, you can move on and press the Serial Monitor button located in the Arduino IDE. If you cannot see it, you should look at the rightmost button of the tool-bar. Any software that can read from the serial port can communicate with the Arduino. Many programming languages will allow one to create programs on the computer that can interact with the serial port.

Motors, Lamps, and others

Every pin in the Arduino board can act as a power energy device that uses about 20 milliamps. You should note that 20 milliamps is not a large current; it is only sufficient to power an LED. If you attempt to power more significant loads such as a motor, the pin will instantly not work, and you risk burning the entire processor.

To power large loads, you need to have an external device that can modify things. An example of this device is the MOSFET transistor. Do not bother much about the name even though it sounds hard. It is an electronic switch that can be powered by applying a voltage to any of its three pins. It resembles something like a switch that we can use at home. In this example, the switch that turns on and off is on the Arduino board.

If you do not know what MOSFET means, it stands for "metal-oxide – semiconductor field-effect transistor." This is a unique type of transistor that operates by the field-effect principle. In short, electricity will go through a semiconductor material when the voltage flows through the Gate pin.

Complex Sensors

Complex sensors refer to sensors which produce a particular kind of information that needs more than just a digitalRead(), or even an analogRead() function. Typically, these are smaller circuits that contain a microcontroller that will preprocess the information.

A few of the complex sensors present consists of accelerometers, ultrasonic rangers, and the infrared ranges.

Digital Functions

So far, you have interacted with several digital functions, and you at least know how to use them. While we have not looked at them in deep, this is the time we want to go deep and examine them so that you can have a better understanding of them.

pinMode()

Before we look and begin to apply other digital functions, it is a good thing to let the Arduino understand the method we want to use the digital I/O pins. To achieve this, we use the above function. The function allows us to define the type of pin that we want to use as well as in which way we are going to use it. It has a straightforward syntax as shown below:

pinMode (pin, state)

Within the parentheses, two values are alienated with a comma. The first value has the pin that we want to set up. This can be a number or a variable that has a value in the range between 0-13 and even A0 –A5.

The second value represents the state that expects the pin to work with the circuit. The state can include two predefined constants. That is the INPUT or OUTPUT. We apply the INPUT in the pinMode() function when we position the pin in a high impedance state. In this mode, the pin is designed to accept an input signal even though it exerts a smaller load on the entire circuit. This is great for one to read sensitive inputs without any impact to the sensor. For a digital input pin, it has sensitivity for two types of values, the "LOW" and "HIGH."

If we apply the OUTPUT, the digital pin assumes a state of "low impedance."
In addition, this can subside a source current given the size of these little chips of 40 milliamps.

digitalWrite()

Once there is a pin that has been identified as the OUTPUT, it allows one to turn the pin on or off using this function. The syntax is shown below:

digitalWrite (pin, state)

This function is used in two statements that have the pin number and the state. It is similar to pinMode(). The value can be between 0-13 and even A0-A5. The next line contains the output state that is similar to the predefined constants. The constants can be HIGH or Low.

The HIGH state of the source current establishes a connection to the +5VDC. LOW represents the default nature of an output pin. This will provide a link to the ground. On the other hand, HIGH switches the circuit on and sets the pin LOW.

digitalRead()

When you have the digital pin configured to the INPUT, it provides the ability to read the state of each pin using this function. The basic syntax includes:

digitalRead(pin)

In the following case, we have to describe the pin number for the particular INPUT pin with the help of a numerical constant or variable. Whatever happens with this type of reading depends a lot on the two separate ways that we can take advantage of this function. The first one is the condition of the pin, and the next one is the application of the function in the variable position.

sensorState = digitalRead(sensorPin);
if (sensorState == HIGH) digitalWrite (ledPin, HIGH);

Reviewing the lines of the code above, we see that the digitalRead() function reads the input pin, the reading is then stored in a variable. The next line uses the if statement to test the variable. Now, if this variable is found equal to HIGH, then the out pin will be HIGH. Still, there is a different way to rewrite the whole code into a single line as shown below:

if(digitalRead(sensorPin) == HIGH) digitalWrite (ledPin, HIGH);

This line of code uses the digitalRead() function in the place of the variable. This means that when the reading of the sensor is equal to HIGH, the rest of the line is executed.

State Changes

Given that we have two conditions that can allow us to read or write digital inputs, and outputs whether low or high, we can use it to determine the change in states. In this case, a pin changes from high to low or even low to high.

If you want to record a change of state on a digital input, no need to explain the exact state of the pin. However, what you do is mark the time at which the pin changes the states.

To achieve this, we have to create a comparison of the pin's current state and the last time we recorded a state. If we search for input and discover that it is low, and the last time it was marked as high, the button is said to be pressed.

The process of identifying these changes is called edge detection. We call it edge detection because our primary interest lies in the precise edge that the state changed from state a to b.

```
#define NOTE_C4    262
#define NOTE_D4    294
#define NOTE_E4    330
#define NOTE_F4    349
#define NOTE_G4    392
#define NOTE_A4    440
#define NOTE_B4    494
#define NOTE_C5    523

const int kPinSpeaker = 9;

void setup()
{
  pinMode(kPinSpeaker, OUTPUT);
}

void loop()
{
  ourTone(NOTE_C4, 500);
  ourTone(NOTE_D4, 500);
  ourTone(NOTE_E4, 500);
  ourTone(NOTE_F4, 500);
  ourTone(NOTE_G4, 500);
  ourTone(NOTE_A4, 500);
  ourTone(NOTE_B4, 500);
  ourTone(NOTE_C5, 500);

  noTone(kPinSpeaker);
  delay(2000);
}

void ourTone(int freq, int duration)
{
  tone(kPinSpeaker, freq, duration);
  delay(duration);
}
```

Chapter 6
Sound

We want to produce some noise. We shall use a Photoresistor and piezo element. We shall make a light-based Theremin.

In this Chapter, you will learn how to make sound using the tone() function as well as how you can calibrate the analog sensors.

The Theremin is a device that creates sound by the movement of a musician's hands near the instrument. The Theremin identifies the place of a performer's hands to the antennas. It achieves this by interpreting the change on the antennas. The antennas connect to the analog circuit that builds sound. A single antenna will manage the frequency of sound and the other antenna controls the volume.

Even though the Arduino cannot perfectly emulate the sounds from this instrument, it can replicate using the tone() function.

This will allow a transducer such as a piezo or speaker to move forward and backward at different speeds.

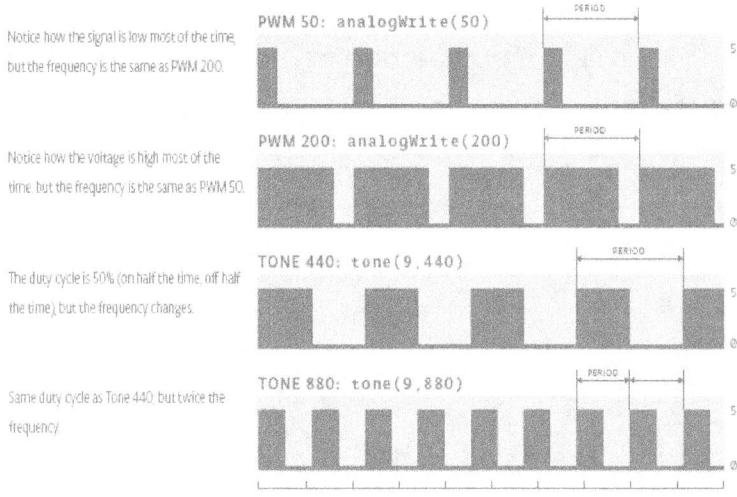

Rather than detecting the capacitance using the Arduino, you will use a Photoresistor to determine the intensity of light. By rubbing your hands over the sensor, you can change the intensity of light that strikes the photo resistor's face. The variation in voltage on the analog pin will show the type of frequency note you need to play.

You will take the photoresistors and connect it to the Arduino using a voltage divider circuit. You should notice that the fixed resistor reduces the lowest end while the brightness of light reduces the high end.

You will not select the limited range, but you will mark the sensor readings so that you can record both the low and high values. This provides one with the ability to adjust the readings of the sensor when you transfer the circuit into a new environment.

Piezo is a small component that will vibrate anytime electricity flows through it. When it moves, it creates sound waves.

Let's build the circuit.

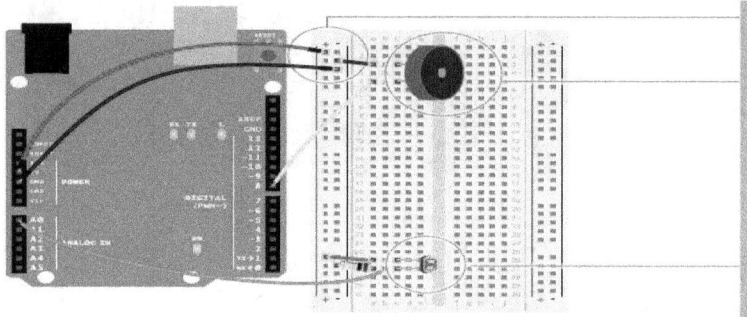

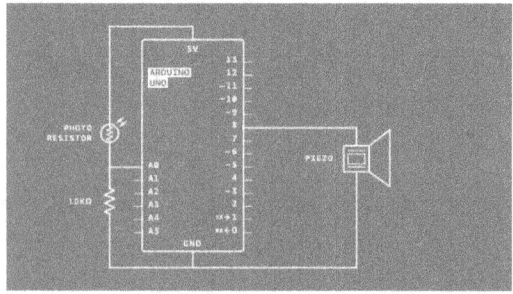

Traditional Theremins could control the volume and frequency of sound. In this example, you will be able to control the frequency. While it is hard to control the volume, it is not something difficult to change the level of voltage that passes through the speaker. What happens when the potentiometer is arranged in series using the piezo and pin 8? What about using a different Photoresistor? Let's find out.

1. Take the breadboard and connect the outer bus to the ground and power.
2. Hold the piezo and connect it to one side of the ground, and the other pin 8 of the Arduino.
3. Insert the photoresistor into the breadboard, and connect one point to the 5V. Next, join the other side to the Arduino's analog pin 0, and to the ground via the resistor.

Let's examine the code:

First, we have to declare a variable to hold the analogRead() value of the Photoresistor. The next thing we declare a variable for high and low values. Set the initial value of the variable sensorLow to 1023 and the variable for the highest value to 0. In the first run of the program, it creates a comparison of the above numbers with the sensor's readings to determine the lowest and highest value.

Identifying the constant for the calibration

Define a constant called ledPin. This should be a signal to show that the sensor has finished calibrating. For this particular example, use the onboard LED connected to pin 13.

Set the digital pin.

In the setup(), update the pinMode() to O and switch on the light.

Use while() to calibrate

The next thing to do is to label both the maximum and minimum values. Allow the while() statement to execute the loop for about 5 seconds. The while() loops will run until when a specific condition is true. In this situation, you will apply the millis() function to find the current time. The millis() reveals the length of time the Arduino has been on from the time it was switched on.

Comparison of sensor values to calibrate

This will take place inside the loop where you will determine the sensor value. In case the value becomes lower than the sensorLow, which was originally 1023, you should update the variable. Likewise, if it is greater than the sensorHigh, which was 0, it is also updated.

Show that the calibration is over

After 5 seconds, the while() loop stops. Remove the LED connected to the pin 13. Calibrate the frequency of your program by using the sensor high and low values.

```
 9    sensorValue = analogRead(A0);
10    if (sensorValue > sensorHigh) {
11      sensorHigh = sensorValue;
12    }
13    if (sensorValue < sensorLow) {
14      sensorLow = sensorValue;
15    }
16  }

17  digitalWrite(ledPin, LOW);
18 }
```

Reading and storing the value of the sensor

While in the loop(), read the value on AO and store it in the sensorValue.

Wrap the sensor value into a certain frequency

Declare a variable called pitch. The value stored in the pitch variable maps from the sensorValue. Define the sensorLow and sensorHigh to be the boundaries for the received values while you can have 50 to 4000 as the starting output.

Play the frequency

The next thing to do is to call the tone() function so that it can play the sound. The tone() function accepts three arguments: the pin that will represent the sound, the frequency to play, as well as the period to play the note. Finally, you can call the delay() function to create a delay of 10 milliseconds so that you create some time for the sound to play.

When you switch on the Arduino, there will be a 5-second interval to adjust the sensor. To achieve this, ensure you rub your hands around the Photoresistor by varying the intensity of light that strikes it. Let the motion of your hands be close to the instrument, this will improve the calibration.

After 5 seconds, calibration is over, and Arduino LED turns off. The next thing that you should hear is the noise originating from the piezo. When the intensity of light that strikes the sensor varies, the frequency of the piezo will also vary.

```
19 void loop() {
20   sensorValue = analogRead(A0);
21   int pitch =
         map(sensorValue, sensorLow, sensorHigh, 50, 4000);


22   tone(8,pitch,20);



23   delay(10);
24 }
```

The map() function defines the pitch as wide, and you can attempt to change the frequencies to determine the one which is perfect for your musical style.

The tone() function works in the same manner as the PWM in the function analogWrite(), but it has one major difference. The analogWrite() has a fixed frequency. However, with the tone(), you will continue to send pulses while you change the rate.

The tone() function allows one to define frequencies when it pulses a piezo or speaker. If you apply sensors into a voltage divider circuit, you will not receive a complete range of values. However, calibrating the sensor allows you to map inputs into a specific field.

Chapter 7
Arduino Shields

An Arduino shield is simply a printed circuit board. You can connect your Arduino so that it extends the functionality in some ways with this extra board. Arduino alone has a few parts. It has the microcontroller, pins, and holes that one can join to the microcontroller.

It has the LED there, right? Then we have the USB port that is used to communicate with the computer. Apart from that, it seems not to have a lot when it comes to the interface. It is important and significant in an advanced system since it helps provide specific functions that are hard to accomplish with the microcontroller.

Therefore, a shield simplifies the process for you. A shield comprises of two things. The hardware that makes up for the little board, a printed circuit that is equivalent to the size of the Arduino and it fits perfectly on the Arduino. The shield is available in the pre-wired state, pre-manufactured board, and you can still purchase it when it is manufactured.

There exist shields that are not manufactured too. This means you can go and purchase the design. It is open-source; you can purchase the design as well as produce it yourself. However, most of the time, you will buy it when it is pre-manufactured.

Therefore, we have the hardware part of the shield and the software part. The software part represents the library that has been linked to the shield. In other words, we use this shield with a collection of libraries, and a set of functions to achieve all the interesting tasks that a shield can deliver.

Well, one might ask, "What are the benefits of the shields?" Why should one use a shield? There is no wiring required. The picture below represents a shield. It is almost like an Ethernet shield.

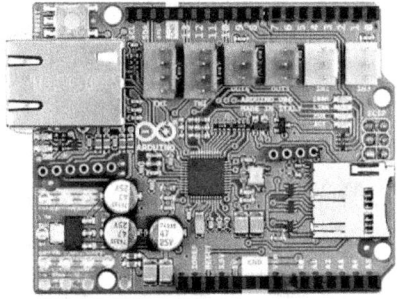

The circuit in the above shield is pre-wired. Thus if you spend a few minutes to review the shield, you will find that it has different components. There are chips that are wired together. That type of wiring should not scare you as a user. You do not necessarily need to do that as a user.

That wiring has been done for you already. It comes when it is pre-wired, so there is no need to be scared of the wiring details. Again, the wiring, which links the rest of the shield and the shield pins to the original Arduino pins, should not worry you. That one is done for you already by placing the shield on top of the Arduino.

If you check at the shield, you will see that it has pins below it, the pins fit into the holes of the Arduino. This means you do not need to be scared of any additional type of wiring. You only pick the board and place it on top of the Arduino, and you will be through with the installation.

That is a beautiful thing, right? There is no extra work needed to build the hardware design of the circuit.

The shield is easy to use because the library functions handle all the complicated details as well as the way to use the device. For instance, in the earlier picture that we saw, we had the Ethernet controller. Inside you will find a chip called the wiznet Ethernet controller. That is the type of complicated chip to use. If you use the chip independently, you will see some data sheet—a thick data sheet that you need to understand.

With the existence of library functions, no need to master any of that. What you will only do is to call them, and there you will establish an Ethernet connection.

Pins

If you look at the picture shown at the beginning of this chapter, you might think that the pins are absent. That is incorrect. The pins are right at the bottom. Look at the Arduino, and you will see a set of holes. The rows of holes are located at the bottom or top, left or right based on your pin orientation.

These shields contain pins at the bottom that go deep to join the holes on top of the Arduino. In other words, you can get the shield, and attach it there, and the wiring will happen automatically.

Even though all the pins are connected, not all the pins are used in communication between the Arduino and the shield. This is something critical especially when you have many shields. Let us assume that you have two shields that you want to stack. If both shields use the same pins for various purposes, then the communication will not be successful.

For instance, if we take one shield which you are using as a digital input and the other pin is applied as the output then here you have a problem. Any time you are going to use multiple shields at the same time, you've got to be careful and note which pins are used for a given shield as well as in which way they are used.

The typical way for a shield to communicate with the Arduino is via the I-2-C. If you have three shields with you, and both communicate via the Trough 12C. They can exactly share similar pins, and there shall be no conflict. If you want, you can combine as many as you would. However, there are certain times when two different shields share the same pin for separate purposes without the presence of any conflict.

Choosing the Arduino Shield

Arduino is a large microcontroller that permits designers to prototype projects. It has a lot of GPIO as well as peripherals. In this section, we look at some factors to consider when looking for an Arduino shield.

The Arduino consists of a sequence of microcontrollers that have a lot of IO connectors as well as the peripherals. However, this might not be enough. You can imagine when an Arduino Uno user would want to combine WI-FI capabilities into their project. They might buy the WI-FI model then link it to the specific GPIO pins before implementing the Uno so that it uses the module. However, Arduino designers had an early thought. They chose to design their boards in such a way where they might combine other boards that have additional features.

The Arduino Shield

Now we know how great the Arduino shield can be, but we don't see how we can go about choosing the best shield. The first thing you should do is to identify the board. The Arduino name covers an entire collection of development boards. Because of the many choices, there are shields that are not compatible with some Arduinos.

Here are a few things to look at:

Pinout

Nearly every type of electronic device has a pinout. This points to where we have the input/output pins placed together with their function. It is essential that the type of shield you use should have the same pinout that your Arduino is using. There are many ways in which you can achieve this, and the list below has different approaches that you can apply.

1. Review the data sheet. It has the type of Arduino boards it works with.
2. Research the shield online to find out what other people have said about.
3. Search for the images of the shield and determine whether the pins share a similar location to the Arduino board.
4. Take the shield physically and put it on top of the board to check whether the pins line up.

The Operating Voltage

While a shield might look compatible because of the pinout, it can still fail to fit with the voltage levels. Specific Arduino boards have 3.3V, and some might use 5V. Because of this, your shield either might destroy the Arduino or be damaged by it. You can identify the levels of voltage by examining the shield's datasheet. This should consist of information such as the I/O voltage, power dissipation, voltage, and current flow. Many shields have been designed with the Arduino Uno in mind; this implies that they are mainly likely to work at 5V. The ARM-based Arduino works at 3.3V.

Libraries

Just because the shield is compatible with the voltage and pin does not imply that the libraries belonging to that shield will run. This might happen because many libraries do not have the Arduino library to identify the I/O pin numbers as well as the hardware application. However, the libraries have direct access to the hardware through the registers that are unique to the ARM or AVR cores. Even though this improves the performance, it also shows that a library written for the Uno might not perform despite the different architectures.

Picking your shield

Now you know some of the technical details to consider an Arduino shield to use in your project. Now is the right time to pick one! To find a shield, identify the reason why you need and the function it will perform. For instance, the Uno does not support internet capability, and something such as the Wi-Fi circuit will be ideal. Below we share with you some of the best four Arduino shields you can purchase and use in your projects.

Best Four Arduino Shields

You have walked into the electronic shop and purchased a complete Arduino starter kit, and not even a single step you have skipped while using the Arduino guides. Still, you have a problem. You need a lot of bobs and bits to fulfill your electronic dream. Fortunately, if you have an Arduino board, you can combine the functionality on top of another one like shields. Shields have been created mainly for the Uno board. As mentioned earlier, shields have the same shape and pin alignment. You plug it on, and you will get an immediate upgrade of the functionality.

The Ethernet shield

It is the formal field from the creators of Arduino. The shield has a way in which it can make your Arduino project to function independently from the computer while maintaining the connection.

Another extra feature this type of board comes with is the MicroSD card slot. Therefore, if your project needs an extensive data file such as the mp3 or video, you can keep it there.

However, before you can buy one, it is important to note that the Ethernet shields are version specific. You might purchase one and fail to fit your board. Therefore, it is essential to find out the version number at the bottom of your board before you can move on to purchase one.

The 4-Relay shield

It is possible to switch on your kettle while you are on the internet. Relays are a critical part for most home automation projects. This is because they can let one switch on and off higher voltage circuits. There is a higher chance that you have come across a relay if you have a starter kit, and you will realize that you need specific components.

With this four-relay shield, you will have said goodbye these problems. With the 4IO pins, you should be able to pull a pin HIGH to turn the corresponding relay. Every relay will deal with about three amps even though you can go with a relay that has low power circuits as a replacement of the on-off switch.

Something that I would like to mention is that if you have let this switch turn on the mains power for your electronic devices, then you should be careful because it is like exposing the wall socket.

The Protoshield

The Protoshield does not perform anything, and that is the reason why it looks empty. However, if you have been using a breadboard to prototype and display your projects, you will have to make it permanent. After that, you can proceed and add more shields on top.

The LCD Screen with 16x Character Display

The reasons to use the LCD screen for your project are apparent. Your Arduino can show messages, but these need about 7 or even more IO pins. This particular type of shield is built with the help of 12C communications bus. This shows that it carries two pins that are related to other parts of the same bus.

Besides the screen, we have a four directional and a select button created. This provides one with an interactive interface that needs to be attached to the PC. If you find that monotone is taking you well, then you can upgrade to this.

Just keep in mind that not all types of shields are stackable. Some will require one to move to the top of the stack since it does not have pins.

Chapter 8
Projects

The Keyboard Instrument

With the help of some buttons, resistors and other devices, you will build a small musical keyboard.

Although you can join a few switches into a digital input and produce different types of tones, in this particular project, you will learn how to build a resistor ladder.

This is a method where you can read some switches by using an analog input. It is an essential technique if you have limited digital data. You will connect some parallel switches into the analog. When you touch each button, there will be a separate voltage level that will flow through the input pin. If you press down two buttons simultaneously, you will find a unique input that depends on the link between the two resistors arranged in parallel. The figure below shows a ladder and five switches.

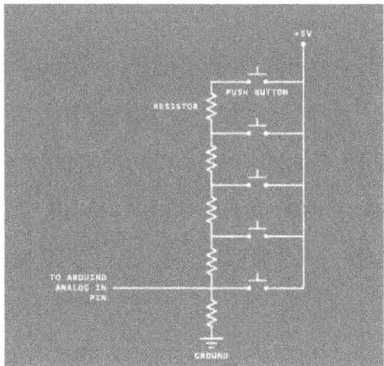

Let's look at the circuit

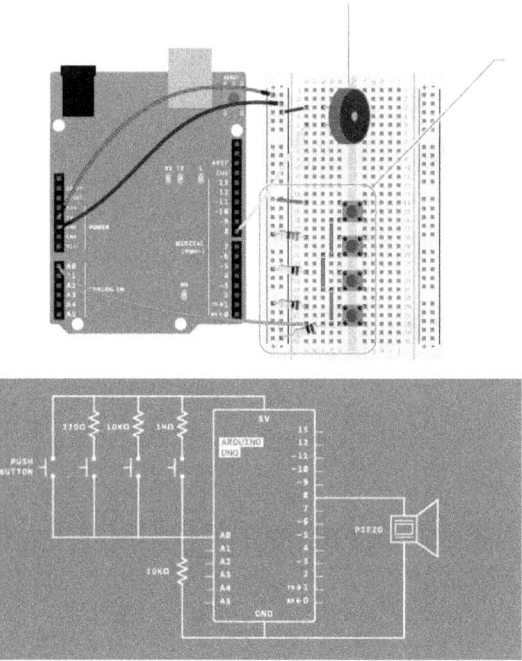

1. First, you have to connect the breadboard to the power and ground like the one we did in the previous examples. Hold one side of the piezo and join it to the ground. Connect the remaining side of the pin to your Arduino.

2. Arrange the switches on the breadboard the same way it has been done in the figure above. The pattern of the resistors and switches going into the analog input is the resistor ladder. Connect the first one straight into the power. Then do the same for the second, third and fourth switches through the respective resistors. Connect the junction to the ground by using a resistor of 10-kilohm and then join it to the analog in 0.

Look for something to enclose your keyboard. Take a small cardboard piece that you can cut to fit your buttons. Mark the keys to help you remember the notes that have been triggered by every key.

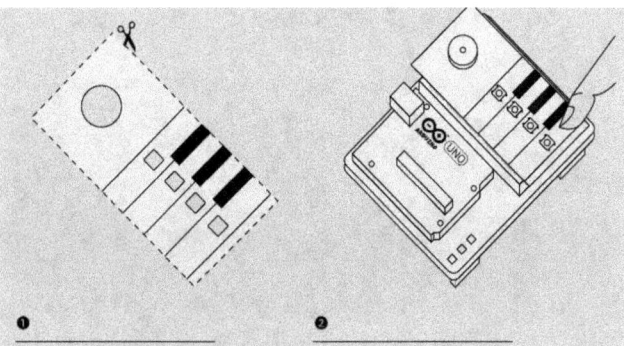

3. Take a small piece of paper that has holes that belong to the four buttons and the piezo. Decorate it so that it can resemble a piano keyboard.

4. Place the paper on top of the buttons and piezo, and enjoy the creation.

The CODE

This program requires one to have different frequency values which you want to play when you touch the buttons. You can start with the frequencies for the
Parts labeled C, D, E, and F. To achieve this, you must create a variable called array.

An array holds different values of similar type; this can be the frequencies of a musical scale. They are a great tool to help one access information fast. If you want to declare an array, do the same way you declare a variable, but make sure the name follows with a pair of square brackets. The elements of the array remain in the curly brackets.

Anyone who would like to change the elements of the array has to first reference the individual elements by listing the name and index of the element. The index is the order by which things will appear in the array. The first index in an array is 0 and the next is 1. This order follows that trend until the last element is reached.

Creating an array of frequencies

Declare an array to store four notes. Make the array global by making the declaration before the start of the setup() function.

In the loop(), create a local variable to hold the reading of pin A0. Given that each switch has a different resistor value that connects to the source of power, there will be unique values. To see the values, use the line below

Serial.println(keyVal)

We have used an if...else statement to help us allocate every value to its particular tone. This program has used random figures for the size of the resistor. Do not use the exact figures in your program because resistors have some errors, this may fail to work in your case.

```
int buttons[6];
// set up an array with 6 integers

int buttons[0] = 2;
// give the first element of the array the value 2
```

```
1  int notes[] = {262,294,330,349};
```

```
2  void setup() {
3    Serial.begin(9600);
4  }
5  void loop() {
6    int keyVal = analogRead(A0);
7    Serial.println(keyVal);

8    if(keyVal == 1023){
9      tone(8, notes[0]);
10   }
```

Play notes that are similar to the analog value

Call the tone() after every if() statement call. The program tells the array to calculate the frequency to play if the value of A0 is similar to the one in the if statements, you can allow the Arduino to play the tone. There is also a possibility that your circuit is noisy and the values can rise when you press the switch. Therefore, it is a good thing to use small values to validate.

If you apply the "&&," look for multiple statements to determine if it is correct. When you press the first button, the notes in the first element will play, touching the second button, the notes in the second element plays and the cycle continues.

To stop the note from playing you use the function noTone(). Just specify as a parameter the pin number you want to stop.

However, in case you have resistors close to one another like in the example program, you should hear sounds originating from the piezo when you press the buttons. If you don't understand, navigate to the serial monitor and make sure that every button is within the range of the notes in the conditional if statement. If you hear a stuttering sound, increase the scale a bit.

Press several buttons simultaneously, and see the type of values that appear in the serial monitor. Use the new values to generate more sounds. Test as many frequencies and expand the musical output.

The tone() function is the best when you want to generate sounds. However, it has some limitations. For instance, the function can only create square waves but not smooth sine waves. Square waves are different from typical waves. While you are about to begin your band, remember that only a single tone can play every time and the function tone interfaces with the analogWrite() on pins 11 and 3.

Note

Finally, remember that arrays are important when you have a similar type of information that you want to classify together. You access arrays using index numbers that point to distinct elements. Resistor ladders provide the right circuit to channel digital signals into a system by inserting into an analog input.

DIGITAL HOURGLASS

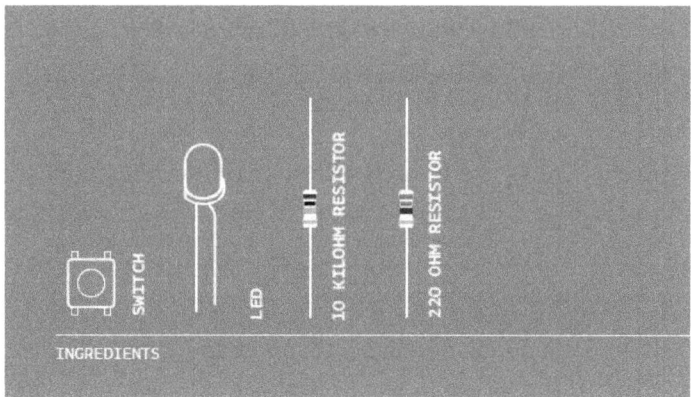

In this project, you are going to learn how to build a digital hourglass that switches on an LED after 10 minutes. This will help you know the time you spend working on your projects.

So far you have seen that when you want to make something happen at a specific time interval, you have to apply the delay() function. This is convenient but at the same time limited in what it can achieve. When the delay() is called, it stops the flow of current based on the time of delay. This means that there is no input and output during the delay. Besides, delays are not the best to use to monitor time. If your goal were to do something after every 20 seconds, a delay of 20 seconds would be very long.

The millis() function comes in to provide a much better solution to this problem. The function will record the time the Arduino has been on.

Up to now, we have declared variables as an int. An int consists of a 16-bit number that contains values in the range "-32,768 and 32,767." That is a large number, but not when the Arduino is making a count of 1000 times a second using the millis() function, in just a few minutes you will be out of space. The long data type can store a 32-bit number.

Given that time cannot run back to produce negative numbers, we declare an unsigned long variable to store millis() time. A data type of unsigned type can only be positive. In addition, an unsigned long can extend to 4,294,967,295. This is sufficient space to store time for even 50 days. So, if you compare the function millis() to a given value, you can tell if a certain amount of time has ended.

So when you rotate your hourglass, a tilt switch will update its state, and that begins a different LED cycle.

The tilt switch operates the same way as a normal switch where it has an on and off sensor. In this project, you will use it as a digital signal. Something unique about tilt switches is the way they determine the orientation. Usually, it contains a small cavity with a metal ball. If adjusted correctly, the ball will roll to one particular side of the cavity and join two leads in the breadboard.

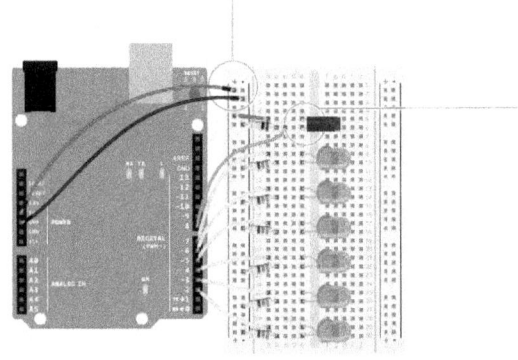

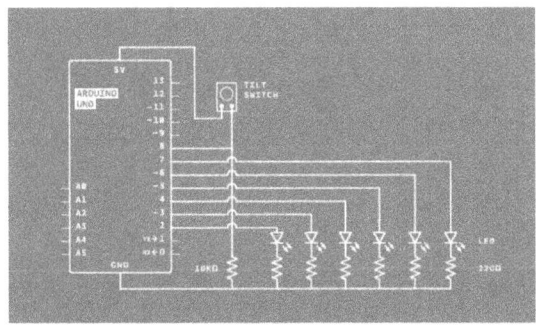

1. The first thing is to connect the power and ground to your breadboard

2. The next thing to do is to join the six LEDs to the digital pins 2-7 through the anode. The LEDs have to be connected to the ground through a resistor.

3. Connect one lead of the tilt switch to the 5V. The remaining part should be connected to the ground.

Create a stand using the cardboard and allow power to flow to the Arduino using a battery. You can build a cover that has some numeric displays close to the lights.

Tilt switches are cheap and affordable components to help one tell the orientation of something. Another example of tilt sensors are the accelerometers. In addition, they are quite expensive. If you are only interested to see whether something is up or down, you should go with the tilt sensors.

The Code

Define a constant

In this project, you will need to have several global variables so that you can have everything work. To begin with, define a constant called switchPin.

Declare a variable to store time

Declare an unsigned long variable. This variable will record the last time the LED changed.

Declare variables to hold inputs and outputs

Define variables for both the switch state and previous switch state. The input and output variables will help make a comparison of the switch's position from one state to the next. Declare a variable called *led*. This will make the next LED to switch on. You can start with pin 2.

Variable declaration showing the interval between two events

The last variable to define is the interval between each LED. This is the long datatype.

Determine the time the program started.

Once the loop() begins, you can find the time the Arduino has been on using the function millis() and place it in a variable called currentTime.

Determine the time that has elapsed since the first loop()

With the help of an if() statement, you need to determine whether time has reached to switch on the LED. Perform some mathematical operations by subtracting the currentTime from the original time and test to see if it is more than the interval variable.

```
1 const int switchPin = 8;

2 unsigned long previousTime = 0;

3 int switchState = 0;
4 int prevSwitchState = 0;

5 int led = 2;

6 long interval = 600000;

7 void setup() {
8    for(int x = 2;x<8;x++){
9        pinMode(x, OUTPUT);
10   }
```

```
11    pinMode(switchPin, INPUT);
12  }

13  void loop(){
14    unsigned long currentTime = millis();


15    if(currentTime - previousTime > interval) {
16      previousTime = currentTime;
```

Switch on the LED, and prepare for the next

The previousTime displays the last time the LED was on. The moment the previousTime is set, switch on the LED and increase the led variable. If you pass the time interval again, the next LED lights up.

Find out whether all the lights are on

Create another if statement in the program to help you determine whether the LED on pin marked 7 is on. Make sure you do not attempt anything with it.

Towards the end of the loop, save the state of the switch in the prevSwitchState, and compare it with the value you receive for the switchState in next loop().

If you are done with programming the board, look at the time in the clock. Once 10 minutes pass, the first LED has to be turned on. After every 10 minutes, a new light will display. After an hour, all the six light will turn on.

```
17    digitalWrite(led, HIGH);
18    led++;


19    if(led == 7){
20    }
21    }

22    switchState = digitalRead(switchPin);


23    if(switchState != prevSwitchState){
24      for(int x = 2;x<8;x++){
25        digitalWrite(x, LOW);
26      }

27      led = 2;
28      previousTime = currentTime;
29    }
```

Chapter 9
Troubleshooting

As you work on your projects, there are specific situations when there will be troubleshooting and debugging.

The more you use Arduino and electronics, the better you become and gain experience. This will end up making the whole process less painful. Don't be frustrated with the problems that you experience. It is much easy than the way it might look at the start.

Since each Arduino project consists of a hardware and software, there are many places to check when things go wrong. Therefore, when debugging, you need to consider these three aspects:

Understanding

You should strive to understand as much as you can in the way the parts you have in your project operate and how they should contribute to the final project. This method will allow you to develop ways in which you can check every component independently.

Simplify and segment

In the ancient Romans, they had what we call divide and rule type of government. You need to break down your project into different components using your intelligence to determine which part of a given function starts and ends.

Exclusion and Certainty

While investigating, you need to check each component individually to ensure that you are sure every component works by itself. You will slowly develop your confidence and note the parts of the project that are doing well and areas that you should fix.

Debugging is the term we refer to the process of fixing errors in software. It was first used by Grace Hopper back in the 1940s. This was the time when computers were majorly electrochemical. It is believed one computer stopped working after an actual insect found its way inside.

However, many of today's bugs are not physical like in this case. Instead, they are virtual and invisible. This means they call for more time and that process can be boring.

Testing the Board

What if the first example of blinking an LED failed to work? That would possibly be frustrating. Let us see what you can do.

Before you throw complains at your project, you need to ascertain that several things are in the right place. This is similar to the way airline pilots follow when they run through a list of things to make sure that there will be no problems when the plane takes off.

The first thing you need to do is to plug your USB into your computer:

- Verify that the PC is on. This might look obvious, but you could forget to turn on your computer. When the green light labeled PWR lights up, this implies that the computer is powering the board. However, if the LED looks faint, then something is wrong with the power. You can change the USB cable and check the computer USB port as well as the Arduino USB plug port on your computer.

- If you are using a new Arduino, the yellow LED labeled L will start to blink.

- Now, if you had been using an external source of power and you have connected an old Arduino. Just ensure that the power supply is inserted in and the jumper labeled as SV1 has been connected to the two pins that are close to the external power supply connector.

Another point you need to note is when you are experiencing problems with some sketches, and you want to verify that the board is working. Open and transfer the blink an LED example in the IDE to the board. The onboard LED has to blink in a regular pattern. If you follow all the above steps, then you should be confident that the Arduino is going to work correctly.

Test the Breadboard Circuit

To test your breadboard circuit, join the board to your breadboard by executing a jumper from the GND and 5V connections right between the negative and positive rails of the breadboard. When the green PWR LED goes off, remove the wires. This shows that there exists an error in the circuit and there is a short circuit. A short-circuit leads to an excessive current which cuts off the current as a mechanism to protect your computer.

However, if you are concerned that you might destroy your computer, remember that most machines have safety mechanisms. Besides, the Arduino board has an independently powered USB hub.

If you have a short circuit, it is essential to apply the divide and conquer approach. What you require to do is to look for each sensor in the project and connect each sensor one at a time.

The first thing that you need to begin with is the power source. Review each part of the circuit to ensure that power flows through it. Work procedurally and perform a single change in every step.

Any time you are in the process of debugging and things do not seem to go well, the best thing to do is to handle everything systematically. This is very important because it will help you fix the problem, and that is why it is crucial that you update one at a time.

In addition, do not forget that each debugging process will stick in your head. You will develop an understanding of some of the things to fix up when you encounter a problem. In addition, after some time you will become an expert at doing it.

IDE problems

There are times when you might experience problems with the Arduino IDE, primarily if you are working on Windows. If you receive an error when you double-click the Arduino Icon or nothing happens, you should attempt to execute the run.bat file, which is another option to start Arduino.

Windows users might again run into a problem if the operating system allocates the COM port to a COM10 or higher number to the Arduino. If this takes place, you can let Windows assign a lower port number to the Arduino.

Look for Help Online

If you find yourself stuck entirely such that you are spending a lot of time trying to debug, it might be time to turn to the community of users at the Arduino forum. One of the best things, when you look for help online, is that you will always find someone ready to assist you if you can describe your problem correctly.

Develop a practice of cutting and pasting things into a search engine and wait to check the results if there is a person who has tried to solve it. Look around to discover a solution, nearly every problem you encounter must have an answer.

You, start with checking the main Arduino website before you can move to the playground. Another critical thing to note is before you begin your project, you should search in the playground for a few lines of code or circuit diagram to help you build your project.

Chapter 10
Make your Project

This is the time for you to make use of your creativity. I hope that as you have been learning the different chapters of this book, you started to develop an idea of the things you can build. One of the best things about microcontrollers is the number of things you can achieve with it.

It is the perfect time to develop your project. In this chapter, we provide you with some of the ideas that you can make. Try to make something from the list below or whatever you have been doing:

- Build a countdown timer which has a speaker, pushbuttons, and LCD
- An alarm which goes on when the temperature goes beyond the limit
- Build a dial which indicates the levels of light
- Develop an alarm clock
- Build a tone playing on the speaker depending on the amount of light or even temperature.
- Type a message into an LCD using a joystick and again playback it on the speaker with the help of a Morse code.

- Build a traffic light simulation using push buttons for the car sensors

Conclusion

When it comes to Arduino, the IDE will help anyone to write the code that will direct your Arduino board on what it can perform. The software is majorly focused on processing. This is a popular IDE and software mostly used by many people. Just like processing, the Arduino has a clear syntax as well as an easy API. This makes it right for prototyping concepts or even teaching people ways in which they can code.

The Arduino functions are a collection of C or C++ functions. The functions can be called while inside the code; this implies that the basic C and C++ functions will still operate in the Arduino. Given that the Arduino board is similar to any other AVR development board, it offers you the opportunity to use the rest of other build tools such as the Makefiles.

Now that in this book you have learned the different things about the Arduino sensors, boards, accessories, and shields, you should be good to translate that theory into a practical project. Pick an Arduino project from the hub and begin to implement it.

If you found some inspiration in reading this book, the next thing I would suggest is to look for more books and read to ensure that you internalize the concepts. Besides just reading other books, make sure you build practical projects, which can further expand your level of understanding.

This is the end of this book, but this should be the beginning of your journey to building bigger projects. Remember: the power of creativity is vested in you; this means you can build any kind of project you want with the help of Arduino. It only takes that leap of faith and courage to start.